EVOLUTION
FACT OR FABLE?

THE CASE AGAINST DARWIN'S BIG IDEA

J. ROBERT KIRK, J.D.

ISBN 978-1-63885-318-3 (Paperback)
ISBN 978-1-63885-319-0 (Digital)

All biblical citations were taken from the
King James Version of the Holy Bible.

Covenant Books
11661 Hwy 707
Murrells Inlet, SC 29576
www.covenantbooks.com

To all those willing to pursue the truth regardless of the cost.

Buy the truth and sell it not; also wisdom,
and instruction, and understanding.
—Proverbs 23:23

CONTENTS

PREFACE

You have been lied to about Darwin's theory of evolution. I am confident in saying this because the lies are everywhere—even biology textbooks are *full* of them. If you have ever looked at a picture of the Darwinian "tree of life" or a series of five creatures with an ape on the left and a modern man on the right, or if you have ever read about a variety of other "icons of evolution," you have been lied to. And, like me, you probably believed the lies.

But if you are like me, you hate being lied to. Truth matters. And when the lie embodies the false idea that human beings are nothing more than a mindless accident of nature and that human life has no purpose, it matters a lot.

I am now sixty-eight years old, but I still have a clear memory of sitting at a library table when I was a sophomore in college at the University of Wisconsin studying for an anthropology exam and thinking about the theory of evolution. Our anthropology professor had taught us that the development of embryos proved evolution's theory of the common descent of all creatures. He told us that as an embryo develops, it actually looks like the creature's evolutionary ancestors. He taught this is why human embryos actually have "gill slits" at one point of development—because eons ago humans shared a common ancestor with fish. We memorized the phrase "ontogeny recapitulates phylogeny"—the idea that biological development tracks evolutionary history.

Neither my professor nor my textbook ever mentioned that this well-known icon of evolution concerning embryos was based entirely on a set of embryo drawings that had been faked in 1866 by a biologist named Ernst Haeckel—which fakery had been well-known for many decades. No one ever mentioned that the parallel lines on the neck of a human embryo are really just folds of tissue, not "gill slits" at all. Nor were we told that even fish embryos don't have "gill slits" at that early stage of development.

Nevertheless, as I sat there thinking about the theory of evolution, I had my doubts. We had been taught that evolution proceeded by random mutations which, if advantageous, somehow accumulated over an uncountable number of tiny steps to produce a fully developed trait that aided the survival or reproductive success of the organism. But I had grown up on a small sheep farm producing lambs to be sold for meat. I knew that our bucks had been selected for their solid "meat type" bodies. But I also knew that these bucks produced lambs that, depending on the character of the mother, often did not look a whole lot like their "meaty" father but more like their slender "dairy type" mothers (not a good thing for us meat producers).

Sitting at that library table, a very specific question crossed my mind. Thinking about a human hand, I imagined that maybe six fingers on a hand might be better than five. Maybe six fingers would somehow aid reproduction or survival. Suppose one person was born with a mutation that produced a small pre-finger bump just next to his right little finger. When he reached the age of reproduction, there would be no reason to think that the mate he chose would also have the same small bump since mutations are completely random, varied, and rare. So what were the chances that this guy would produce children with that same small bump? If the bump was a "recessive" trait, it seemed there would be no chance at all. Even if it was somehow "dominant," what was the chance that his child or grandchild would actually have the same random and rare mutation that would somehow make that little bump slightly bigger and be one of many steps closer to becoming a functional finger? And even if he did, what was the chance that little bump would be preserved by natural selection—since it was still far from being anything useful, much less a semi-functional finger?

In other words, how did the evolution of brand-new structures *actually happen*—not just in theory, but in actual reality?

Since all my imagining was asking questions I could not answer quickly and was distracting me from studying for my exam, I simply set it aside. For more than twenty years. I set it all aside for more than twenty years.

When I graduated from the University of Wisconsin-Madison in 1975, I totally believed that the theory of evolution was as well established as the theory of gravity. When I graduated from Harvard Law School in 1978, I believed the very same thing. It wasn't until my wife and I had precious identical twin daughters in 1992 that anything changed. For the very first time, I needed to know the actual truth about the biggest ideas in life so that I could teach our precious daughters the truth. Ideas like "Where did people come from?" and "Does life have a purpose?"

So for the first time in my life, I undertook a study of the theory of evolution for myself and for our precious daughters—not to earn a grade, but to know the truth.

The first of many books I read about evolution was Michael Denton's *Evolution: A Theory in Crisis*.[1] By the time I finished reading fifteen chapters of scientific analysis of Darwin's theory, my world was shaking dramatically. Denton, an Australian biochemist, closed his text with the statement: "Ultimately the Darwinian theory of evolution is no more nor less than the great cosmogenic myth of the twentieth century."[2] And Denton had persuaded me that his conclusion was probably right.

The more I read about the theory of evolution, the more confident I became that Denton was right. I read many books, including a reprint of *Origin of Species* itself.[3] I also read *The Blind Watchmaker*, considered to be one of the most powerful modern arguments in favor of Darwin's theory.[4] And I have continued to study new books about Darwinian evolution through the end of 2020.

More than 160 years after the publication of *Origin*, the evidence against Darwin's biggest idea is now overwhelming. This

[1] Michael Denton, *Evolution: A Theory in Crisis, New Developments in Science are Challenging Orthodox Darwinism* (Bethesda, MD: Adler & Adler, 1985).

[2] Ibid., 358.

[3] Charles Darwin, *Origin of Species* (New York: Random House, 1979, originally published in 1859 by J. Murray, London, under the title, *On the Origin of Species by Means of Natural Selection*).

[4] Richard Dawkins, *The Blind Watchmaker: Why the Evidence of Evolution Reveals a Universe Without Design* (New York, NY: W.W. Norton & Co., 1996).

should not be surprising since Darwin and his contemporaries knew almost nothing about microbiology, microphysiology, heredity, genetics, DNA, or a wide variety of other dimensions of bio knowledge that have grown explosively in recent decades. As well-known biochemist and Darwin critic Michael Behe has written in his most recent book, *Darwin Devolves* (2019),

> The firm conclusion I've drawn over the past decades is this: despite occasional questions and bumps along the road, the greater the progress of science, the more deeply into life design can be seen to extend.[5]

Notice I am referring to the weight of the scientific evidence which now contradicts the theory of evolution, not merely the weight of expert opinions. The arguments against Darwin's biggest idea are overwhelming in power but definitely not in popularity—indeed, such arguments are hardly ever heard. It is almost impossible these days to be a successful academic in any life science and be opposed to Darwin's big idea. Opposing Darwin is more likely than anything else to get you tossed out of academia altogether.

Happily, I am not an academic. I am a retired trial lawyer with experience proving and disproving facts. And I'm old enough to still believe that facts matter much more than opinions, conclusory statements or storytelling. Those things can be entertaining, but they are woefully inadequate if you're thinking seriously about ideas as important as "Why are we all here?"

So if you care about real evidence when it comes to life's big ideas and you don't have time to read the many books I've read concerning the theory of evolution, I invite you to join me for a brief summary of the most powerful evidence showing that Darwin's biggest idea was wrong. My goal is to explain this evidence in a way that

---

[5] Michael Behe, *Darwin Devolves: The New Science About DNA That Challenges Evolution* (New York: HarperCollins, 2019), 10.

is easy both to understand and to remember so you can explain it to others.

And once you are so equipped, I sincerely hope that you do explain the mythology of Darwinism to others because every person deserves to know that he or she is much more than a mindless accident.

CHAPTER 1

FIRST THINGS FIRST

The Problem of Definitions

Discussions of Darwin's theory of evolution are often very confusing because different people use the word *evolution* to mean very different things. Darwin presented more than one idea in *Origin of Species.* Deciding which part of his ideas one is talking about is extremely important because one part of what he said is obviously true while another part of what he said is completely contrary to the evidence.

As an attorney responsible for proving and disproving facts, I learned over time that this job is most difficult by far when the thing to be proven is partly true and partly false. Since this is the case with Darwin's theory of evolution, it is very important to understand at the outset the exact boundary line between these two parts of the theory.

Michael Denton describes the two parts of Darwin's theory this way:

> In his book Darwin is actually presenting two related but quite distinct theories. The first, which has sometimes been called the "special theory," is relatively conservative and restricted in scope and merely proposes that new races and species arise in nature by the agency of natural selection, thus the complete title of his book: *The Origin of Species by Means of Natural Selection or*

the Preservation of Favoured Races in the Struggle for Life. The second theory, which is often called the "general theory," is far more radical. It makes the claim that the "special theory" applies universally; and hence that the appearance of all the manifold diversity of life on Earth can be explained by a simple extrapolation of the processes which bring about relatively trivial changes such as those seen [in finches] on the Galapagos Islands. This "general theory" is what most people think of when they refer to evolution theory.[6]

Darwin's "special theory" is known to me as "micro-evolution" or what I call Darwin's "little idea." It is objectively true—almost by definition. Darwin had observed that populations of creatures exhibit slight variations among individuals, that these individuals compete among themselves for resources, and that those individuals with variations most well-adapted to their surroundings survive best and reproduce most. This biological reality came to be known as "survival of the fittest." Who can argue with this? It explains why a mutant virus can survive a medicine that killed most of its kind prior to the mutation or why after a couple million years of mutations among the finches on the Galapagos Islands, Darwin saw several different varieties of finches with various small differences, such as different shaped beaks. Different beaks, but they were all still finches.[7]

Darwin's "general theory" is what is known to me as "macro-evolution" or what I call Darwin's "big idea." Darwin's big idea was that if variation, competition, and natural selection could explain changes in the shape of beaks among finches on some islands in the Pacific, that very same mechanism could, simply by extrapolation, explain all variation among all creatures everywhere. Obviously, this was a

---

6 Denton, *Evolution*, 44.
7 Please don't make the mistake of thinking that if life on earth is billions of years old, then Darwin's big idea must be true. As I discuss in chapter 6, this is not the case.

very, very big idea that could not be built on Darwin's observations but only his imaginations. But Darwin's imaginations concerning "common descent" were so great that he actually speculated not only that all of the millions of types of plants and animals on earth had descended from only a handful of life-forms, but that all plants and all animals had both descended from some unknown single original form of life. He wrote:

> It may be asked how far I extend the doctrine of modification of species…I believe that animals have descended from at most only four or five progenitors, and plants from an equal or lesser number. Analogy would lead me one step further…Therefore I should infer from analogy that probably all the organic beings which have ever lived on this earth have descended from some one primordial form, into which life was first breathed.[8]

And so students are universally taught that some unspecified "primordial soup" somehow gave rise to some original form of life—perhaps some single-celled creature—which somehow turned into a multicelled creature, which after some untold expanse of time turned into creatures with backbones beginning with a fish that later turned into an amphibian, that still later gave rise to reptiles and mammals, and still later some of the reptiles turned into birds. While these hypothetical transitions are never, ever described in detail, the general lesson that all known life "has evolved" from some simpler form of life has been taught so often and for so long that its truth, at least among the "intelligentsia," simply goes without saying.

Even as late as 2014, honest Darwinists have had to admit that there is "little choice but to resort to *our more-or-less informed imagination* to produce the historical narratives that are the ultimate goal of our studies of animal evolution. *[O]ur imagination* is the only tool

---

[8] Darwin, *Origin*, 454–455.

that can braid the fragmentary evidence into a seamless historical narrative that relates the *what, how,* and *why*" of evolution.[9]

So proponents of evolution often equate micro and macroevolution even though the two ideas are entirely different. They present examples of microevolution for the purpose of proving the truth of macroevolution even though they do no such thing. And evolutionists know full well that equating the two ideas is based entirely on presumption, not evidence. One well-known Darwin apologist—the one who actually coined the terms "micro" and "macroevolution"— put it this way:

> There is no way toward an understanding of the mechanics of macroevolutionary changes which require time on a geological scale, other than through a full comprehension of the microevolutionary processes observable within the span of a human lifetime and often controlled by man's will. *For this reason we are compelled at the present level of knowledge reluctantly to put a sign of equality between the mechanisms of macro- and microevolution, and proceeding on this assumption,* to push our investigations as far ahead as this working hypothesis will permit.[10]

But the mechanisms of macro and microevolution are definitely not equal, and the best research of the twenty-first century has made this more and more clear. As Michael Denton has explained in his most recent book, there is currently "widespread dissatisfaction with the micro- to macro-extrapolation"—so much so that the conclusion that natural selection "can't be the whole story about how pheno-

---

[9]   Ronald A. Jenner, "Macroevolution of animal body plans: Is there science after the tree?" *BioScience* 64 (2014), 653–664. (Emphasis added.)

[10] Theodosius Dobzhansky, *Genetics and the Origin of Species* (New York: Columbia University Press, 1937), 12. (Emphasis added.)

types [large groups of organisms] evolve…isn't seriously in dispute these days."[11]

As we review the evidence, we will presume the truth of micro-evolution even as we discuss the evidence contradicting Darwin's big idea of macroevolution. The very most recent evidence suggests not only that macroevolution did not happen but that it actually could not have happened.

"Natural Selection" Is Itself a Somewhat Deceptive Term

As noted, the definition of evolution includes the idea of survival of the fittest. The term most commonly used to describe this part of the process is the phrase "natural selection."

But it is important to note that this term is somewhat deceptive. This is because the word "selection" implies the involvement of some sort of mind or plan which is doing the "selecting" based on some preexisting goal. But this is not Darwin's idea at all. Darwin's big idea contends that both the mutations producing variation and natural selection itself are completely *mindless* processes. The only "selection" that is going on is that individuals with certain attributes are surviving more frequently than individuals without those attributes.

In the year 2000, Phillip Johnson wrote a book called *The Wedge of Truth* which contained a whole chapter dealing with the problem of the complexity of life.[12] The chapter was called "The Information Quandary." The problem is that macroevolution requires new genetic information to build new organs or other biological systems, yet mutation plus natural selection cannot produce such new information. In passing, Johnson described a big problem with the phrase "natural selection."

---

[11] Michael Denton, *Evolution: Still a Theory in Crisis in Crisis* (Seattle, WA: Discovery Institute Press, 2016), 26, quoting Jerry Fodor and Massimo Piattelli-Palmarini, *What Darwin Got Wrong* (London: Profile Books Ltd., 2010), 77.

[12] Phillip E. Johnson, *The Wedge of Truth: Splitting the Foundations of Naturalism* (Madison, WI: InterVarsity Press, 2000).

Simply as a matter of logic, it is absurd to describe natural selection as information generating. Natural selection does not actually "select," much less create. The term is nothing more than a misleading metaphor that merely stands for the proposition that early death or sterility is not necessarily random. Some creatures may survive to reproduce because they have some advantageous quality (like the ability to run faster than their peers when pursued by predators), but that quality has to *be there* already before natural selection can favor it. Death doesn't do any creating, whether it is random or not.[13]

In an earlier book, Johnson had concluded that biologists "assume that natural selection has great creative power, not because that power can be demonstrated but because there is no better naturalistic alternative."[14]

The lesson concerning a phrase such as "natural selection" is that words are very important. Be careful to focus on the actual reality concerning a matter being described rather than merely the connotation of a carefully chosen phrase. A very big part of the problem of the evolution story is that it is based on imaginations rather than physical reality. In other words, it is fiction.

Binomial Nomenclature

Since we are talking about *Origin of Species*, it is useful to say a few words about the meaning of the word "species." The sixth edition of Darwin's *Origin of Species* contained a glossary that defined many technical terms but, oddly, he did not provide a definition for the term "species."

---

[13] Ibid., 59. (Emphasis original.)
[14] Phillip Johnson, *Defeating Darwinism by Opening Minds* (Downers Grove, Illinois: Intervarsity Press, 1997), 56.

A century before Darwin, a Swedish botanist named Carolus Linnaeus had formulated a system for assigning a unique name to every different group of organisms on earth. His system assigned category names to the very biggest groups, as well as smaller and smaller groups, all the way down to the individual unique species.

Most people are aware that human beings are also known as "Homo sapiens." This is an example of Linnaean "binomial nomenclature." This term means the use of two words to identify a unique species. The word *species* generally means organisms of the same form or, more specifically, the largest category of organisms that can reproduce with each other.

From top to bottom—or broadest to narrowest—the categories Linnaeus defined for human beings are as follows:

kingdom	animal
phylum (phyla)	cordate
class	mammal
order	primate
family	hominid
genus (genera)	homo
species	sapiens

Our discussion of issues frequently will refer to these categories because they provide an easy way to describe things like the amount of change that has been observed in a population of creatures. For example, when we talk about the most recent evidence concerning the genetic variations found in creatures such as Darwin's Galapagos finches, we will see that such variations can explain changes observed in the form of organisms at the level of species, or perhaps even genus, but not at the level of family or higher.

It is important to keep in mind that classification of organisms into particular Linnaean groups—especially at the lowest levels—is a very subjective process that depends greatly on the biases of the one doing the classification. Historically, it involved a researcher simply deciding whether the particular form of an organism looked more like those in one category or another. Even in most recent days when the genetic maps of organisms are considered, a judgment must still be made about whether or not a certain amount of genetic difference justifies declaring the existence of a new species.

And subjectivity often is at play even with topics that seem to be controlled by mathematical objectivity. One quick example of this relates to the favorite claim of Darwinists that the entire set of human genes, or "genome," is 96 percent the same as the genome of a chimpanzee. This number is often cited as "proof positive" that humans and chimps share a common ancestor—in other words, that Darwin's big idea is true. But what is often left unsaid is that the human genome consists of 3.2 billion base pairs—meaning that it is 6.4 billion letters long. Four percent of that is a difference of 128,000,000 base pairs or 256,000,000 letters.

So is a difference of 128 million base pairs or 256 million letters a small difference or a big difference? Sounds pretty big to me, but you be the judge. The point is that even statements that sound objective, or even mathematical—"*only* a 4 percent difference"—can represent a conclusion that in truth is quite subjective.

Evidence Versus Imagination

It is one thing for a person to have an idea. It is quite a different thing for that person to have evidence in support of that idea.

In law, the American standard of due process embodies fundamental rules of procedure that recognize this important difference. A fact witness at trial in court may only testify to things that he or she personally saw or heard—that is, what he or she observed with his or her own senses. The witness may not merely repeat what he was told by others (hearsay) or make conclusory statements unsupported by actual, personal observations. The witness is never allowed to tell the jury, "I'm sure the defendant is guilty" or "Ten very smart people have told me the defendant is guilty." Instead, the witness is only allowed to be heard if he or she can say something like, "I saw the defendant stab the victim in the chest." Observation, yes. Mere conclusion, gossip, or opinion? No.

Darwin had a very big idea, but he had very little evidence in support of that big idea. When you read *Origin of Species,* you see that it is loaded with indefinite speculative words, such as "could" or "maybe" or "possibly" describing imaginations or guesses rather than observations. Here are just a few examples:

- Concerning the possible evolution of the human eye:
  - "[W]e *ought in imagination* to take..."
  - "Further *we must suppose...*"
  - "*[M]ay we not believe...*"[15]

- Concerning the lack of fossil evidence:
  - "I can answer...only *on the supposition that...*"
  - The "duration [of rock formations], *I am inclined to believe*, has been shorter...*"
  - These are "questions on which *we are confessedly ignorant*; nor do we know how ignorant we are."[16]

- Concerning the action of natural selection:
  - "I can under such circumstances *see no reason to doubt* that..."
  - "Some of its young *would probably* inherit..."
  - "[A]nother process *might* commence...hence we *may believe* that..."[17]

Just as it was in Darwin's famous book, modern discussions of the theory of evolution are loaded with imaginations rather than evidence. For example, evolution advocates are very fond of saying things like "the scientific community agrees that apes and humans

---

[15] Darwin, *Origin*, 219. (Emphasis added.)
[16] Ibid., 439–440. (Emphasis added.)
[17] Ibid., 138, 139 and 141. (Emphasis added.)

shared a common ancestor." Statements such as this have almost no value for many reasons:

- This sentence merely states a conclusion without describing in any way the evidence, if any, on which it is based.
- The claim invites you to adopt the "popular" position which allegedly is the position of the "experts"—two favorite methods for convincing people they should forget the common sense notion that conclusions such as this should be based on evidence rather than a show of hands by the "experts."
- The history of science tells us unequivocally that even if every biologist in the world actually signed his or her name to the proposition that apes and humans shared a common ancestor, their agreement would provide no evidence at all that the proposition is actually true. For a period of centuries during the Middle Ages, virtually every scientist on earth believed in geo-centrism—the idea that the earth is at the center of our solar system. It was not until the 1500s that an astronomer and mathematician named Copernicus advocated the idea of helio-centrism—the opposite idea that the earth and all other planets revolve around the sun. And it took probably another century for that true fact to become widely accepted among the "experts."

My dear readers, the point is that every single one of you is a true expert in what really matters, which is *common sense*. Common sense is based on actual life experience and it is an extremely important, valuable thing—most especially in a world chock-full of false ideas. It's properly called "common" sense because it is shared by all thinking people.

You use common sense every day to succeed in life. So if a man knocks at your door, says he is from the Internal Revenue Service, says he is on official business and wants to ask you a few questions, you probably ask to see some sort of IRS identification badge and probably write down the name and number on that badge. But if the first question he asks is "What is your credit card number, expiration date, and security code?" you probably send him packing. Because people with common sense don't just believe everything they are told. They know that certain things just don't make sense. This kind of common sense represents very great wisdom and is extremely valuable.

Both evolutionists and their critics have urged people to develop sensitive "baloney detectors," which is simply a pet name for the logic and investigative skill that are part of common sense. In his book *Defeating Darwinism by Opening Minds*, Phillip Johnson included a chapter titled "Tuning Up Your Baloney Detector."[18] Johnson agreed with astronomer Carl Sagan on the great importance of developing a baloney detector, but Johnson advocated that the well-developed detector should also be pointed at the claims of science itself—and especially the claims of biologists advocating Darwin's theory of macroevolution.

Johnson argued that a well-developed baloney detector would be able to spot and untangle the most favorite strategies of baloney salesmen. These are things like the "appeal to authority," discussed above—where you are urged to accept a conclusion because the supposed "experts" all agree that it is true. Or "the big lie," where a complete falsehood is declared to be true with very great bravado. Or the "half truth," where the truth of microevolution is cited as evidence for the truth of macroevolution—which actually is a completely different idea.

Be encouraged, even when thinking about complicated "scientific" ideas. You can understand and sort the evidence well because you have common sense. Don't be distracted by whatever propaganda you may have heard. Just consider the evidence—not the loudly repeated slogans, but the evidence.

It has been said that Darwin's big idea is one of the few scientific ideas on earth that is believed by almost all scientists but disbelieved by a majority of the public. The good reason for this fact is that most

---

[18] Johnson, *Defeating Darwinism*, 37–52.

people have common sense. As you read, you will see that disbelieving Darwin's big idea makes a whole lot more sense than believing it because his idea actually is contradicted by the facts.

Darwin Himself Was Much More Honest than Most

It is fair to say that Darwin himself was much more honest about his theory than many of his modern advocates. Darwin actually seemed to have his doubts about his big idea. We don't have to guess about this because Darwin actually admitted these doubts in writing—sometimes in *Origin of Species* itself. For example, concerning the absence of fossil evidence of the many transitional forms predicted by his big idea, he wrote:

> If numerous species, belonging to the same genera or families, have really started into life all at once [as the fossil record showed at the time], *the fact would be fatal* to the theory of descent with slow modification through natural selection.[19]

> To the question why we do not find [fossil] records of these vast primordial periods, I can give *no satisfactory answer.*[20]

> That the geological record is imperfect all will admit; but that it is *imperfect to the degree which I require, few will be inclined to admit.*[21]

> The case [concerning the lack of fossils] at present must remain inexplicable; and may be truly

---

[19] Darwin, *Origin*, 309. (Emphasis added.)
[20] Ibid., 313. (Emphasis added.)
[21] Ibid., 440. (Emphasis added.)

urged as *a valid argument against the views here entertained.*[22]

By contrast, modern proponents of Darwin's big idea almost never acknowledge any problems with his theory. On the contrary, they are often willing to make statements in support of it that are most extreme. For example, one contemporary paleontologist has said concerning the transition of one type of insect into another—despite a complete lack of solid evidence—that the "macroevolutionary transition from one body form to another with a completely different number of segments and appendages is *a very easy process.*"[23]

Often such advocates resort to mocking skeptics, even expressing anger or disgust with anyone who has the nerve to doubt their contentions. For example, world famous Darwin apologist Richard Dawkins has written, "No serious biologist doubts the fact that evolution has happened, nor that all living creatures are cousins of one another."[24] He has also sneered that "those who seek to deny the truth of evolution…[are]backwoodsmen."[25] But he is most famous for saying, "It is absolutely safe to say that if you meet somebody who claims not to believe in evolution, that person is *ignorant, stupid or insane* (or wicked, but I'd rather not consider that)."[26]

My brief review of the lack of evidence supporting Darwin's big idea will not engage in such name-calling or hostility. However, I mention the problem in case you head off to the internet (or elsewhere) to dig into more detailed evidence on a particular topic. Expect Darwin's doubters to be ridiculed. Expect their work to be mocked. Wikipedia is one of the worst—no doubt because it is openly editable and the great majority of biologists owe their financial future to the continuing good name of Darwin's big idea.

---

[22] Ibid., 314. (Emphasis added.)
[23] Donald R. Prothero, *Evolution: What the Fossils Say and Why It Matters* (New York: Columbia University Press, 2007), 194. (Emphasis added.)
[24] Dawkins, *The Blind Watchmaker*, 287.
[25] Ibid., Introduction at x.
[26] Richard Dawkins, "Review of Blueprints: Solving the Mystery of Evolution" *New York Times* (April 9, 1989). (Emphasis added.)

But be encouraged. If you decide in advance to simply ignore unsubstantiated or hostile statements, you can find the truth. It's not easy, but you can do it.

Think Like a Detective

No, we are not all as clever or observant as Sherlock Holmes. But relying on common sense, we can all apply just a few rules to the process of how we think about evidence to make it much more likely that we will find the truth. This is what detectives do. They don't just accept at "face value" everything they see or hear. They *second-guess everything*. They know that people sometimes have motives that lead them to twist the facts. They know the difference between opinions or conclusions and objective facts. And they ask the right questions.

As a lawyer who has spent many, many hours trying to prove facts, experience has taught me the great power of two of the most important questions that a detective should always ask. These questions do not require "expertise" as much as they require common knowledge about how things in life usually work. They recognize that life is extremely complicated, so things almost never exist in complete isolation but are almost always closely related to a number of other things. These two important questions are:

1. If the thing I am trying to prove actually happened, *what else should also have happened?* In other words, if the fact I'm trying to prove is true, *what else should also be true?*
2. If the thing I am trying to prove actually happened, *what else should* not *have happened?* In other words, if the fact I'm trying to prove is true, *what else should be false?*

Let me illustrate how this can work by asking you to imagine yourself as a member of a jury asked to decide whether a particular defendant is guilty of having shot a man from a very long distance away—the kind of distance that usually requires the expert skill of a sniper.

Question 1: If this allegation is true, what else should also be true? It should be true that the defendant was known to have been in the same town as the victim around the time of the shooting. It should be true that he is known to have owned a gun with a scope or had access to one. Even better, it should be true that he had some training in long-distance marksmanship.

Question 2: And what should be false? It should *not* be true that the defendant was on the other side of the world on the day of the shooting. Or that he was known to be deathly afraid of guns and had never fired one. Or that the shooting victim was the defendant's beloved brother.

You get the picture. Some things make sense and some things don't.

As noted above, Darwin himself was honest enough to admit that if his big idea was true that every living thing on earth had descended from a small number of living things, then a series of other things that people did not yet know must also be true. And if these things turned out not to be true, then his big idea was in big trouble. Like any good detective, Darwin also knew that if his big idea were true, then certain other things must be false. And if these other things turned out to be true, then his big idea would be in even more trouble.

So before we sort through the evidence, let's consider how these two great questions actually apply to Darwin's big idea. I will ask the two questions any good detective should ask and then give a series of answers to each. Many of those answers will be taken from Darwin's own writings.

- Question 1: If Darwin's big idea were true that every living thing on earth descended from some original unspecified form of life in an infinitely long series of tiny adaptations caused by random mutations and preserved by natural selection, *then what else must also be true?*

1. There must be lots of fossil evidence of transitional forms.[27]
2. There must be biological evidence showing how one form of organism might have changed, bit by bit, into a different form.
3. The age of life on earth must be long enough for Darwin's big idea to have had enough time to have occurred.[28]
4. The most common examples or "icons" used to illustrate Darwin's big idea should all be based on solid evidence.

- Question 2: If Darwin's big idea were true that every living thing on earth descended from some original unspecified form of life in an infinitely long series of tiny adaptations caused by random mutations and preserved by natural selection, *then what else must* not *be true?*

  1. No organ exists that could not have developed by an infinite series of gradual tiny steps.[29]

---

[27] Darwin, *Origin,* 292. "[S]o must the number of intermediate varieties, which have formerly existed on the earth, be truly enormous."

[28] Darwin actually had no idea about either the number of years life had existed on earth or how many years it may have taken for various forms to have evolved. For example, concerning the development of a part of the evolutionary "tree of life," he could not say whether such development had taken a thousand generations or a hundred million generations (Darwin, *Origin,* 160–161).

[29] "If it could be demonstrated that any complex organ existed, which could not possibly have been formed by numerous, successive, slight modifications, my theory would absolutely break down." (Darwin, *Origin,* 219).

2. Nothing can be true about any creature that does not benefit that creature's survival and reproduction.[30]
3. No complex organism exists which does not have much less complicated ancestors.[31]

As we will see as we consider the evidence, it turns out concerning the four things that should be true if Darwin's big idea is true—these things are actually all false. And concerning the three things that must be false if Darwin's big idea is true—these things are actually all true. In other words, after more than 150 years of gathering evidence on questions like those even Darwin himself said must be answered, the evidence tells us that Darwin's big idea is false.

---

[30] "Natural selection will never produce in a being anything injurious to itself, for natural selection acts solely by and for the good of each." (Darwin, *Origin*, 229).

[31] "Natura non facit saltum [Nature does not make jumps]. This canon, if we look only to the present inhabitants of the world, is not strictly correct, but if we include all those of past times, it must by my theory be strictly true."

"As natural selection acts solely by accumulating slight, successive, favourable variations, it can produce no great or sudden modification; it can act only by very short and slow steps." (Darwin, *Origin*, 233 and 444).

CHAPTER 2

DARWIN'S "TREE OF LIFE" IS IMAGINARY

Sometime in your life you have probably seen an illustration of Darwin's "tree of life" similar to the one in figure 1. The tree is a metaphor depicting Darwin's idea that the beginning or root of all organisms on earth consisted of no more than a handful of the most primitive forms of life which, over eons of time, evolved into all the forms we see on earth today. Primitive life at the root, modern life at the twigs. Universal common descent depicted in one, simple picture.

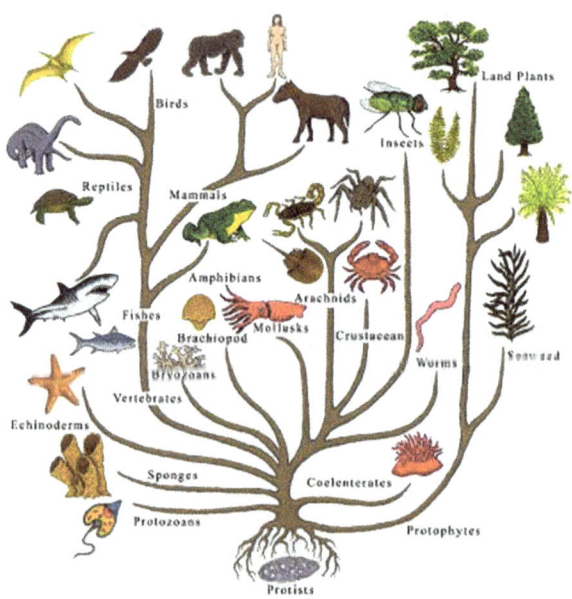

Figure 1. An example of an evolutionary "tree of life." The word "protists" below the roots of the tree refers to an original life-form that is neither animal, plant, or fungus. This original form is thought to be something from a most primitive kingdom.

Darwin himself approved of the tree metaphor, as follows:

> The affinities of all beings of the same class have sometimes been represented by a great tree. I believe this simile largely speaks the truth. The green and budding twigs may represent existing species; and those produced during each former year may represent the long succession of extinct species. At each period of growth all the growing twigs have tried to branch out on all sides, and to overtop and kill the surrounding twigs and branches, in the same manner as species and groups of species have tried to overmaster other species in the great battle for life. The limbs divided into great branches, and these into lesser and lesser branches, were themselves once, when the tree was small, budding twigs; and this connexion [sic] of the former and present buds by ramifying branches may well represent the classification of all extinct and living species in groups subordinate to groups. Of the many twigs which flourished when the tree was a mere bush, only two or three, now grown into great branches, yet survive and bear all the other branches; so with the species which lived during long-past geological periods, very few now have living and modified descendants.[32]

But like many other things concerning the theory of evolution, modern depictions of Darwin's tree of life go far beyond Darwin's own illustration of the tree. Modern illustrations teach that all organisms that ever lived were all the descendants of some single original form. Darwin wrote about this same idea, but his comparable illustration was of only one small part of the imaginary tree—in effect,

---

[32] Darwin, *Origin*, 171.

from some small horizontal branch high in the crown up to some of the highest twigs. Darwin's illustration is shown in figure 2.[33]

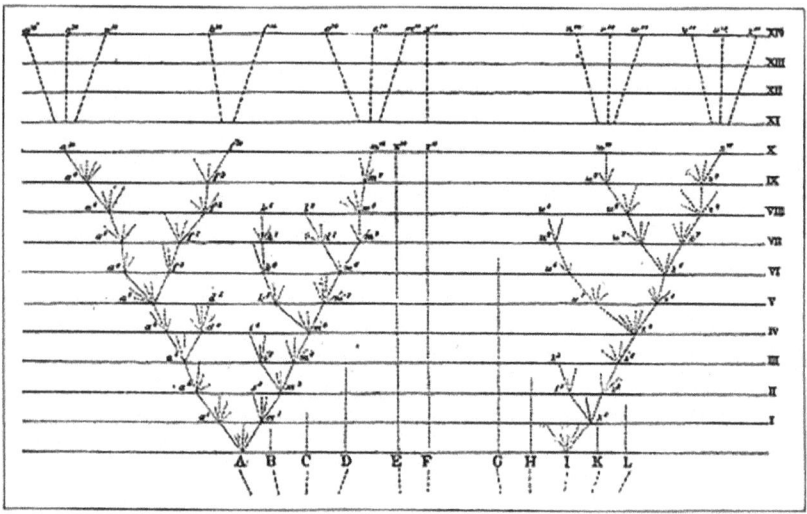

Figure 2. From Darwin, C., Origin of Species, pages 160–161.

Darwin's horizontal lines represented the passage of time, but Darwin had no idea how much time. He admitted that the intervals between the horizontal lines might each represent a thousand generations, ten thousand generations, or perhaps even a hundred million generations.[34] But his point was simply to imagine a general pattern for how a few related species might, over time, become slightly different forms that either go extinct or go on to change even more—eventually becoming completely different types of organisms.

Regardless of the details, Darwin was confident that the record of life on earth would show an almost countless number of extinct and existing forms laid out in a continuous pattern of transition from ancient to modern forms. "On this doctrine of the extermination of *an infinitude of connecting links*, between the living and extinct inhabitants of the world, and at each successive period between the

---

[33] Ibid., 160–161.
[34] Ibid., 162 and 167.

extinct and still older species, why is not every geological formation charged with such links?"[35] Darwin could only imagine that such links would be found.

The fatal problem with all this is that, 150 years later, there is still almost no evidence to support what Darwin imagined. If the tree icon was an accurate depiction of how life actually developed, there should be many, many fossils that testify to the existence of Darwin's "infinitude" of transitional forms. But there aren't. Instead, biologists are happy to simply line up a series of creatures possessing some vague general similarity of form and ask you to imagine that one kind might have evolved into another. For example, the favorite ape to human series seen in figure 3 is not an artist's sketch based on actual fossils but merely an artist's imagination of what transitional forms between apes and humans might have looked like.

Figure 3. From ape to man, based not on fossils but imaginary intermediate creatures.

---

[35] Ibid., 438. (Emphasis added.)

When Darwin wrote *Origin of Species*, very little work with fossils (paleontology) had ever been done. Denton writes that

> in Darwin's day only a tiny fraction of all fossil
> bearing strata had been examined and the number of professional paleontologists could practically be counted on two hands...So vast has been
> the expansion of paleontological activity over
> the past one hundred years that probably 99.9%
> of all paleontological work has been carried out
> since 1860.[36]

As of Darwin's time, the oldest fossils that had been discovered were from the Cambrian period—which occurred roughly five hundred to six hundred million years ago. They showed evidence of all the major phyla of creatures existing that long ago as being fully formed. No simpler transitional ancestors were yet known. But Darwin knew that paleontology was in its infancy, and he was sure that evidence of the transitional ancestors would be found in the future.

We are now in that future and we know that he was wrong. One modern author put it this way:

> The known fossil record fails to demonstrate a
> single example of phyletic (gradual) evolution
> accomplishing a major morphologic transition
> and hence offers no evidence that the gradualistic
> model can be valid.[37]

Indeed, the fossil record shows quite the opposite: a metaphor much better depicted by a "lawn-of-life" rather than a "tree-of-life." We know this based on some very extensive evidence.

---

[36] Denton, *Evolution,* 160.
[37] S. Stanley, *Macroevolution* (San Francisco: W.H. Freeman and Co., 1979), 39.

Ardent evolutionists want to insist that many forms of life must have existed that will never be found because, for example, they were too small or had soft bodies not easily preserved in fossil form. Of course, this excuse completely begs the question of why there are no fossils of more simple ancestors for creatures from the most readily fossilized groups, such as the crustaceans.

> What is even more interesting is that the evidence for Darwinian macroevolutionary transformations is most conspicuously absent just where the fossil evidence is most plentiful—among marine invertebrates. (The animals are plentiful as fossils because they are so frequently covered in sediment upon death, whereas land animals are exposed to scavengers and to the elements.) If the theory were true, and if the correct explanation for the difficulty in finding ancestors were the incompleteness of the fossil record, then the evidence for macroevolutionary transitions would be most plentiful where the record is most complete.[38]

But just as with all other groups, the transitional ancestor fossil cupboard for marine invertebrates is also bare.

Perhaps even more striking is the fact that we now have fossils of *bacteria and single-celled organisms* more than *three billion years old*. And we have abundant examples of completely soft-bodied specimens that "reveal not only their outlines but sometimes even internal organs such as the intestines or muscles."[39] So the notion that ancient

---

[38] Johnson, *Defeating Darwinism*, 60.
[39] J. Wells, *Icons of Evolution: Science or Myth?* 44, quoting Simon Conway Morris, *The Crucible of Creation*, 28.

ancestors are missing from the fossil record because they were too small or too delicate is "now recognized as incorrect."[40]

Accordingly, the matter of when complex life emerged and how it emerged should no longer be in doubt. Essentially, it happened all at once with no long history of gradual divergence as suggested by the "tree of life."

> It is still, as it was in Darwin's day, overwhelmingly true that the first representatives of all the major classes of organisms known to biology are already highly characteristic of their class when they make their initial appearance in the fossil record.[41]

Even the great Darwinian apologist Richard Dawkins is forced to confess that "it is as though [these organisms] were just planted there, without any evolutionary history."[42]

For the animal kingdom, the undisputed fact is that life literally exploded onto the scene during what is now known as the "Cambrian explosion"—an event that took only a few million years to complete. In geologic terms, it was almost instantaneous—a geologic blink of the eye. Of eighteen primary phyla of animals, twelve first appeared during the Cambrian period, with the other six first appearing just before or just after. Jonathan Wells has said that the "fossil evidence is so strong, and the event so dramatic that [the Cambrian explosion has become] 'biology's big bang.'"[43]

---

[40] J. William Schopf, "The early evolution of life: Solution to Darwin's dilemma, *Trends in Ecology and Evolution* 9 (1994), 375–377.

[41] Denton, *Evolution*, 162.

[42] Dawkins, *The Blind Watchmaker*, 229.

[43] Jonathan Wells, *Icons of Evolution: Science or Myth?* 37.

Comparing plants to the Cambrian explosion for animals, Denton explained:

> The story is the same for plants. Again, the first representatives of each major group appear in the fossil record already highly specialized and highly characteristic of the group to which they belong. Perhaps one of the most abrupt arrivals of any plant group in the fossil record is the appearance of the angiosperms [seed-bearing plants] in the era known to geologists as the Cretaceous. Like the sudden appearance of the first animal groups in the Cambrian rocks, the sudden appearance of the angiosperms is a persistent anomaly which has resisted all attempts at explanation since Darwin's time. The sudden origin of the angiosperms puzzled him. In a letter to Hooker he wrote: "Nothing is more extraordinary in the history of the Vegetable Kingdom, as it seems to me, than the *apparently* very sudden or abrupt development of the higher plants."[44]

Wells depicts the Cambrian explosion of animal life in a simple chart showing the first appearance of eighteen phyla almost simultaneously. The eighteen major groups of animals are shown across the top represented by the letters "a" through "r." The geologic time periods are shown in the column on the left. Virtually all of these major groups first appeared during the Cambrian era. Just three phyla appeared slightly before the Cambrian period, and just three appeared slightly later.[45] The animal kingdom has almost no representation at all during the Precambrian period.

---

[44] Denton, *Evolution,* 163. (Emphasis original.)

[45] Wells, *Icons,* 40.

	a	b	c	d	e	f	g	h	i	j	k	l	m	n	o	p	q	r
Recent																		
Permian																		
Carboniferous																		
Devonian																		
Silurian																		
Ordovician																		
Cambrian																		
Precambrian																		

Figure 4. Shaded boxes show the first appearance and continuation of eighteen phyla of the animal kingdom. Taken from Wells, Icons, page 40. (Used by permission of Regnery Publishing.)

The sudden appearance in the fossil record of both fully formed plants and animals is completely inconsistent with Darwin's big idea. Darwin was absolutely convinced that the "canon" declaring that nature does not make jumps "must be strictly true."[46]

But the fossil record proves that massive universal "jumps" are exactly what nature *has* made, both in the plant and animal kingdoms. The very biggest groups within the animal kingdom—phyla

---

[46] Darwin, *Origin*, 233.

and classes—appeared fully formed right from the start. Precambrian fossils consist mostly of single-celled organisms. The few Precambrian multicellular organisms found in the fossil record are so strange that it is not even clear whether they are animals at all or deserve a classification kingdom of their own.[47]

In short, "the now well-documented Precambrian fossil record does not provide anything like the long history of gradual divergence required by Darwin's theory."[48] Understanding this alone, people of common sense cannot still believe in Darwin's big idea.

Remember, Darwin was very well aware that the sudden appearance of life without transitional ancestors was a very serious problem for his big idea. He wrote, "There is another…difficulty which is much graver. I allude to the manner in which numbers of species of the same group, suddenly appear in the lowest known fossiliferous rocks."[49] Darwin even admitted, "If numerous species, belonging to the same genera or families, have really started into life all at once, *the fact would be fatal to the theory* of descent with slow modification through natural selection."[50]

In fact, groups far larger than just genera or families started into life all at once. And this fact is indeed fatal to Darwin's big idea.

---

[47] Wells, *Icons*, 37–38.
[48] Ibid., 38.
[49] Darwin, *Origin*, 312.
[50] Ibid., 309. (Emphasis added.)

CHAPTER 3

DARWIN'S "INFINITUDE" OF TRANSITIONAL FORMS DOESN'T EXIST

Chapter 2 shows that virtually all the major types of plants and animals suddenly appeared on earth fully formed and looking nothing like the handful of original super-simple forms of life imagined by Darwin. And there are no intermediaries. In and of itself, this proves that a huge proportion of the "infinitude" of transitional forms predicted by Darwin does not exist.

But his problem is actually much worse than this. Let's entirely set aside the state of the fossil record during the eons of time prior to the sudden emergence of both plants and animals on earth. What about the countless transitional forms that should exist connecting the most ancient known ancestors of every group of plants and animals with their modern counterparts? Where are all those transitional forms?

To best judge what Darwin's big idea predicted about how the fossil evidence should look, let's review again what Darwin himself actually said:

- "[W]hy, if species have descended from other species by insensibly fine gradations, do we *not everywhere see innumerable transitional forms?*"[51]
- "So that the number of intermediate and transitional links, between all living and extinct species, *must have been incon-*

---

[51] Darwin, *Origin*, 205. (Emphasis added.)

ceivably great. But assuredly, if this theory be true, such have lived upon this earth."[52]

- *"[S]o must the number of intermediate varieties,* which have formerly existed on the earth, *be truly enormous."*[53]

- "As on the theory of natural selection *an interminable number of intermediate forms must have existed,* linking together all the species in each group by gradations as fine as our present varieties…"[54]

- "The mind cannot possibly grasp the full meaning of the term of a hundred million years; it cannot add up and perceive the full effects of *many slight variations, accumulated during an almost infinite number of generations."*[55]

- "On this doctrine of the extermination of *an infinitude of connecting links,* between the living and extinct inhabitants of the world, and at each successive period between the extinct and still older species…"[56]

- "[D]uring these vast, yet quite unknown, periods of time, *the world swarmed with living creatures."*[57]

- "As natural selection acts solely by accumulating *slight, successive, favourable variations,* it can produce no great or sudden modification; it can act *only by very short and slow steps."*[58]

So if, as we know, the most ancient ancestors of all animals and plants first appeared on earth beginning about five hundred to six hundred million years ago, how many transitional forms should we expect to find connecting them with their modern descendants that we see on earth today? Darwin used words like "infinitude," "interminable," "inconceivably great," and "truly enormous" to give us a

---

[52] Ibid., 293. (Emphasis added.)
[53] Ibid., 292. (Emphasis added.)
[54] Ibid., 438. (Emphasis added.)
[55] Ibid., 453. (Emphasis added.)
[56] Ibid., 438. (Emphasis added.)
[57] Ibid., 313. (Emphasis added.)
[58] Ibid., 444. (Emphasis added.)

sense of the number to expect. So it is easy to see that the expected number is not merely a few. We should be expecting to see a very great many indeed.

Here's how I think of it. Imagine your dining room table is the surface of the earth. Take a big saltshaker and shake out salt all over your table—lots of it. Keep shaking until the table begins to be covered with salt crystals. That's what the fossil record should look like. Each salt crystal represents a different species of organism, randomly spaced from its closest relatives by random mutations and natural selection.

Another way of saying this is that, if Darwin's big idea were true, we should not see any groupings of creatures that all share a number of defining characteristics (such as reptiles, birds, or mammals) not shared by others outside the group. But, in fact, that is virtually all we see. Michel Denton has put it this way:

> If evolution has occurred as conceived by Darwin, invariant taxa-defining novelties, not led up to via long sequences of transitional forms from some antecedent structure, *should not exist*. But exist they do![59]

So when Darwin looked at the actual fossil record, what did he see? What he saw was little round conical piles of salt crystals—that is, species—with large empty spaces in between the piles. Each pile represented a type of organism showing slight variations within the type but no obvious transitional forms between the types. As noted above, Darwin knew this was a big problem for his theory, but he was confident it would be solved over time. He had a very active imagination. He imagined that as the surface of the earth was excavated more and more, fossils of transitional forms would be found

---

[59] Denton, *Evolution: Still a Theory in Crisis in Crisis*, 53–54. (Emphasis original.)

that would spread out those little round piles so that they actually touched each other.

After more than 150 years of digging, is that what has happened? Absolutely not. What actually has happened is the opposite. Almost without exception, as "new forms" have been discovered, they invariably relate to forms already known—so the existing piles of types grow bigger and the spaces in between the piles remain empty. Michael Denton summarized the situation this way:

> All the new forms of life which have been uncovered by paleontology invariably relate, whether closely or distantly, as sister species to already known forms and must therefore be placed peripherally in any hypothetical evolutionary tree.
>
> The virtual complete absence of intermediate and ancestral forms from the fossil record is today recognized widely by many leading paleontologists as one of its most striking characteristics...[60]

While many paleontologists have known the truth, the public has not. In 1991, Phillip Johnson recognized, "Just about everyone who took a college biology course during the last sixty years or so has been led to believe that the fossil record was a bulwark of support for the classic Darwinian thesis, not a liability that had to be explained away."[61] And Niles Eldredge, one of the world's most famous paleontologists, has admitted, "We paleontologists have said that the his-

---

[60] Denton, *Evolution*, 165.
[61] Phillip Johnson, *Darwin on Trial* (Downers Grove, Illinois: Intervarsity Press, 1991), 58.

tory of life supports [the story of gradual adaptive change], *all the while really knowing that it does not.*"[62]

A "Missing Link" Is Meaningless Unless You Can See the Whole Chain

Because Darwin's expected "missing links" are virtually nonexistent, any report of such a find almost always receives immediate worldwide attention. One of the most notorious such reports was in 1912 when an amateur paleontologist announced that he had found a prehuman specimen in a gravel pit in Piltdown, England, which came to be known as "Piltdown Man." It wasn't until 1953 that Piltdown man was proven to be a forgery—the intentional assemblage of a modern chemically treated orangutan jaw with a very old yet still modern human skull. The fact that it took forty years for a fairly obvious forgery to be noticed—the teeth of the orangutan had actually been filed down to look more human—taught the important lesson that scientists tend to see in fossils what they want to see.[63]

Another early "missing link" was found in 1861. Biologists have long thought that birds evolved from some form of reptile—perhaps even from a dinosaur. So excitement was great when just two years after *Origin of Species* was published, a fossil called *Archaeopteryx* was discovered and soon declared to be a "missing link" between modern birds and some primitive form of reptile. *Archaeopteryx* had some of the characteristics of reptiles (teeth, long bony tail, and claws) and some of the characteristics of birds (bipedal, wings, and feathers). Thus, it was believed to be a transitional form between these two groups of vertebrates.

---

[62] Ibid., 59. (Emphasis added.)
[63] Wells, *Icons*, 217–219.

Figure 5. Fossil impression of Archaeopteryx on the left; artist rendition on the right.

Archaeopteryx quickly became an icon of evolution, and it remains so to this day. But like all other icons of evolution, the truth is not so simple. Michael Denton wrote:

> No doubt it can be argued that *Archaeopteryx* hints of a reptilian ancestry but surely hints do not provide a sufficient basis upon which to secure the concept of the continuity of nature. Moreover, there is no question that this archaic bird is not led up by a series of transitional forms from an ordinary terrestrial reptile through a number of gliding types with increasingly developed feathers until the avian condition is reached.[64]

We know that *Archaeopteryx* appears to have had fully developed flight feathers and a birdlike braincase. Because many of its soft-tissue characteristics remain unknown, Denton asks, "[M]ight it not have been as avian as any other bird in all important anatomical and physiological characteristics?"[65]

---

[64] Denton, *Evolution*, 176.
[65] Ibid., 178.

But even if *Archaeopteryx* was a bird, it now appears that it was *not* the ancestor of modern birds. Wells cites the research on point:

> [T]here are too many structural differences between *Archaeopteryx* and modern birds for the latter to be descendants of the former. In 1985 University of Kansas paleontologist Larry Martin wrote: "*Archaeopteryx* is not ancestral of any group of modern birds."...And in 1996 paleontologist Mark Norell, of the American Museum of Natural History in New York, called *Archaeopteryx* "a very important fossil," but added that most paleontologists now believe it is not a direct ancestor of modern birds.[66]

Similarly, the chief science writer for the journal *Nature* wrote in 1999:

> Once upon a time, *Archaeopteryx* stood alone as the earliest fossil bird. Its uniqueness made it an icon, conferring on it the status of an ancestor... [But other evidence] shows that *Archaeopteryx* is just another dinosaur with feathers.[67]

The bottom line on all this is that big conclusions should be based on a widespread pattern of evidence—an entire chain of links—rather than a single "missing link." Denton puts it this way:

> To demonstrate that the great divisions of nature were really bridged by transitional forms in the past, it is not sufficient to find in the fossil record one or two types of organisms of doubtful affinity which might be placed on skeletal grounds in

---

[66] Wells, *Icons*, 116.

[67] Henry Gee, *In Search of Deep Time* (New York: The Free Press, 1999), 195–197.

a relatively intermediate position between other groups.[68]

The fundamental problem in explaining the gaps in terms of an insufficient search or in terms of the imperfection of the record is their systematic character—the fact that there are fewer transitional species between the major divisions than between the minor. Between Eohippus and the modern horse (a minor division) we have dozens of transitional species, while between a primitive land mammal and a whale (major division) we have none. And this rule applies universally throughout the living kingdom to all types of organisms, both those that are poor candidates for fossilization such as insects and those which are ideal, like mollusks. But this is the *exact reverse* of what is required by evolution.[69]

In *Evolution: A Theory in Crisis*, I claimed that empirical discontinuities seem to coincide invariably with a major conceptual discontinuity in envisaging how they might have come about—a coincidence which strongly reinforces the notion that the gaps are real and not mere sampling errors. I have no reason to retract that claim. On the contrary, thirty years on, I believe it is more secure than ever.[70]

To help us imagine just how many transitional forms ought to exist if Darwin's big idea were true, Denton considered the specifics

---

[68] Denton, *Evolution,* 177.
[69] Ibid., 191–192. (Emphasis original.)
[70] Denton, *Evolution: Still a Theory in Crisis*, 119.

of one of the many thousands of gaps in the actual fossil record—the gap between land mammals and whales.

> Considering how trivial the differences in mor-phology usually are between well-defined species today, such as rat-mouse, fox-dog, and taking into account all the modifications necessary to convert a land mammal into a whale—forelimb modifications, the evolution of tail flukes, the streamlining, reduction of hind limbs, modifica-tions of skull to bring nostrils to the top of the head, modification of trachea, modifications of behavior patterns, specialized nipples so that the young could feed underwater (a complete list would be enormous)—one is inclined to think in terms of possibly hundreds, even thousands, of transitional species on the most direct path between a hypothetical land ancestor and the common ancestor of modern whales.[71]

Yet do those hundreds or thousands of expected transitional forms exist? Or even some small fraction of them? No. But Darwinists still insist on pointing to a couple of vaguely similar creatures and demanding we conclude that an ancestral relationship must exist.

The Links between Land-Mammals and Whales Are Still Missing

Michael Denton is quoted a few paragraphs above saying that the number of known transitional forms between land mammals and whales was "none" at the time his book was published in 1985. As of that time, Denton presented illustrations of skeletons to show the form of a most primitive whale and a certain land-based creature often nominated by evolutionists as being one of the whale's closest

---

[71] Denton, *Evolution*, 174.

terrestrial relatives. One can quickly see that these are two entirely different creatures.

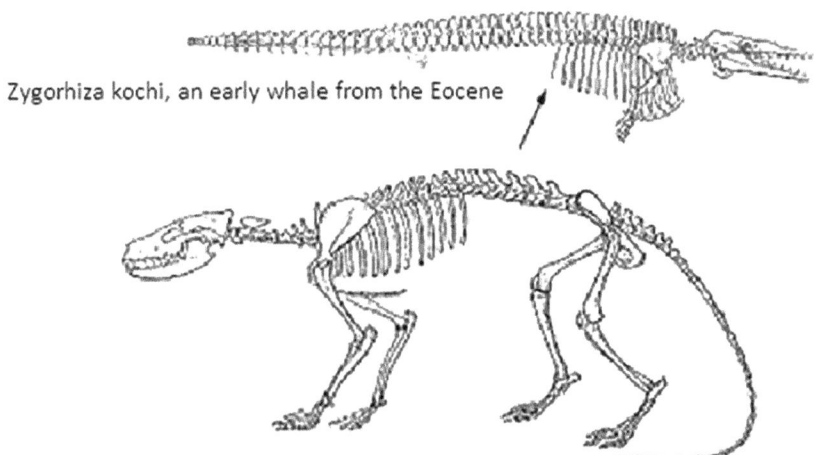

Figure 6. Land mammal "ancestor" below early whale skeleton at top. Taken from Denton, M., Evolution, page 170. (Copyright the American Museum of Natural History, 1951. Used by permission.)

About eight years after Denton noted the absence of transitional forms between land mammals and whales, a new transitional candidate emerged. In 1993, about 80 percent of the fossilized skeleton of a creature dubbed *Ambulocetus*—"walking whale"—was found in Pakistan. The connection between this creature and modern whales was relatively minor—mainly the similarity of a bone of the inner ear called the involucrum. Nevertheless, the experts proclaimed with confidence that the *Ambulocetus* shown in figure 7 must be an ancestor of the ancient whale shown at the top of figure 6.

Figure 7. Skeleton of Ambulocetus—the "walking whale."

Indeed, as a few more skeletons of walking whales were discovered in the 1990s, a growing crowd of evolutionists began to declare to the world that the series of fossils linking land-based mammals to whales was perhaps the best series of fossils between major groups ever discovered. One of the world's most famous advocates of evolution, the late Harvard professor Stephen Jay Gould, went so far as to call it "the sweetest series of transitional fossils an evolutionist could ever hope to find" and "a triumph in the history of paleontology."[72]

What we were not told is an entire series of facts showing that these so-called walking whales were not ancestors of whales at all.

In 2017, Jonathan Wells wrote a book called *Zombie Science: More Icons of Evolution*.[73] One of the chapters of his book was called "Walking Whales," in which Wells debunks this newest icon of evolution—the supposed land mammal to ocean whale transition.

Referring to the creatures strung together as the supposed ancestors of the modern whale, Wells' review of the science led him to the

[72] Stephen Jay Gould, "Hooking Leviathan by its past," Natural History 103 (May 1994), 8–14.

[73] Jonathan Wells, *Zombie Science: More Icons of Evolution* (Seattle, WA: Discovery Institute Press, 2017).

conclusion that "none of these animals were really whales, or even close."[74] In support of this conclusion, Wells noted the following:

- The presence of that inner ear bone called the involucrum thought to connect the *Ambulocetus* to whales can no longer be used for this purpose because an involucrum has been found in a creature called the *Indohyus* which belongs to an entirely different order of mammals—a group including pigs, sheep and cattle—having nothing to do with whales.[75]

- Many of the so-called walking whales are often presented in artist's sketches that make them appear to be swimming when, in fact, they were land-based animals who spent most of their time walking on land. Other of the supposed whale ancestors have distinguishing characteristics that demonstrate they are neither ancestors nor descendants of others in the supposed line.[76]

- As Michael Denton had noted in 1985, in order for a walking whale to become an actual whale it would need transformations in a variety of major systems, including complex organ systems needed for swimming, deep diving, breathing, reproduction and many other features.[77] Wells agrees that such changes would have required "at least hundreds or thousands of mutations…to explain how 'walking whales' evolved into modern cetaceans [whales]."[78] Wells reports that the estimated ages of the fossils used by Darwinists

---

[74] Ibid., 103.

[75] Ibid.

[76] Ibid., 102–103.

[77] See the discussion in chapter 4 below of the amazingly complex system for cooling whale testicles.

[78] Ibid., 112.

would require that the walking whales became modern whales in only about eight million years. But Wells says that the science of mutation rates teaches that "fixing *just two* mutations in [whale ancestors] would take millions of years longer than the time available in the fossil record."[79] At that rate, fixing hundreds or thousands of mutations in the time available would have been a complete impossibility.

- In 2016, this problem of timing got even worse because of the discovery of a modern whale fossil which was several million years older than what had been thought to be the date of the first appearance of modern whales. If accurate, this fossil means that there would have been virtually no time for the hundreds or thousands of mutations required to transform a walking whale into a modern whale.[80]

Wells concludes his discussion of walking whales with a lesson that applies quite often in studying the claims of those advocating Darwin's big idea.

So "the sweetest series of transitional fossils an evolutionist could ever hope to find" is not so sweet after all. It quickly sours with a little additional digging.

With enough imagination anyone can invent a story about how land animals evolved into whales. But an imaginative story is not empirical science. When the materialistic story of whale

---

[79] Wells, *Zombie Science*, 113. (Emphasis added.)
[80] Ibid., 113–114.

evolution ignores inconvenient evidence, it is zombie science.[81]

If Darwin's big idea were true, there should be at least dozens, if not hundreds or thousands, of clear-cut transitional forms of life that are easily understood to connect simpler groups of creatures with more well-developed groups. Instead, there are literally only a couple. And even those are suspect in important ways.[82] Years before his recent unfounded optimism based on fossils of the so-called walking whales, the late Harvard professor Steven Jay Gould called these gaping holes in the fossil record "the trade secret of paleontology."[83]

Remember that Darwin said, "[T]he distinctness of specific forms and their not being blended together by innumerable transitional links, *is a very obvious difficulty.*"[84] And again, "The case [concerning the lack of fossils] at present must remain inexplicable; and may be truly urged as *a valid argument against the views here entertained.*"[85]

A valid argument indeed.

---

[81] Ibid., 114.

[82] Please note that the transitional forms cited by Denton between Eohippus ("Dawn Horse") and the modern horse are evidence of microevolution, not evidence of Darwin's big idea.

[83] S.J. Gould, *The Panda's Thumb* (New York and London: W.W. Norton and Co., 1980), 181.

[84] Darwin, *Origin*, 291. (Emphasis added.)

[85] Ibid., 314. (Emphasis added.)

CHAPTER 4

YOU CAN'T GET THERE FROM HERE

Darwin was very concerned with the problem of how random mutations and natural selection might have created "organs of extreme
perfection," such as the human eye.[86] He began the sixth chapter of
his famous book devoted to "Difficulties on Theory" with the admission that "[s]ome of [the difficulties] are so grave that to this day I
can never reflect on them without being staggered."[87] Concerning
the problem of complex organs, Darwin confessed:

> To suppose that the eye, with all its inimitable
> contrivances for adjusting the focus to differ-
> ent distances, for admitting different amounts
> of light, and for the correction of spherical and
> chromatic aberration, could have been formed by
> natural selection, seems, I freely confess, absurd
> in the highest possible degree.[88]
>
> If it could be demonstrated that any complex
> organ existed, which could not possibly have
> been formed by numerous, successive, slight
> modifications, my theory would absolutely break
> down. But I can find out no such case.[89]

---

[86] Darwin, *Origin*, 217.
[87] Ibid., 205.
[88] Ibid., 217.
[89] Ibid., 219.

One reason Darwin could find no such case was his willingness to rely on guesswork rather than observations in imagining how something as miraculous as the human eye might have come about. His words are literally a series of imaginations. For example:

> With these facts, here far too briefly and imperfectly given, which show that there is much graduated diversity in the eyes of living crustaceans, and bearing in mind how small the number of living animals is in proportion to those which have become extinct, *I can see no very great difficulty (not more than in the case of many other structures) in believing* that natural selection has converted the simple apparatus of an optic nerve merely coated with pigment and invested by transparent membrane, into an optical instrument as perfect as is possessed by any member of the great Articulate class.[90]

But "I can see no very great difficulty" is storytelling, not science. Darwin's long, run-on sentence contains multiple guesses, presumptions, and unsupported conclusions—but very few observations. Darwin's presumption that the optic nerve is a "simple apparatus" is itself breathtakingly naive. How exactly does this "simple" nerve tissue actually transmit a visual image from the eye to the brain in a form that allows the creature to see? Darwin was probably comfortable making statements such as this because in a day before microbiology, no one had any idea how such things actually occurred—so wild guesses were the norm.

Happily, wild guesses are no longer the norm. In 1996, biochemist Michael Behe published a fascinating book called *Darwin's Black Box: The Biochemical Challenge to Evolution*.[91] Unlike Darwin,

---

[90] Ibid., 218. (Emphasis added.)
[91] M.J. Behe, *Darwin's Black Box: The Biochemical Challenge to Evolution* (New York, NY: The Free Press, 1996).

Dr. Behe could see a vast number of organs in nature that could not have evolved by numerous, successive, and slight modifications as required by Darwin's big idea. A major portion of his book explains why several specific examples of widespread biologic realities could not possibly have evolved as Darwin imagined.

Behe's big idea is that in an age before microbiology, Darwin saw much of biology as a "black box." Darwin had lots of ideas about the general form of organisms, but he knew almost nothing about how life worked at the molecular level. As a biochemist, Behe insists that we cannot understand much about biology today unless we understand it at a molecular level.

> Modern science has learned that, ultimately, life is a molecular phenomenon: All organisms are made of molecules that act as the nuts and bolts, gears and pulleys of biological systems. Certainly there are complex biological features (such as the circulation of blood) that emerge at higher levels, but the gritty details of life are the province of biomolecules. Therefore the science of biochemistry, which studies those molecules, has as its mission the exploration of the very foundation of life.[92]

Of course, Darwin could not even imagine the molecular level of life. If he had been able to do so, he would have never made statements that we now know to have been wildly ignorant.[93]

---

[92] Behe, *Darwin's Black Box*, x.

[93] It is hard to overstate the ignorance of Darwin's assertion that anything about vision is "simple." In his book *Darwin's Black Box*, Michael Behe provided a very brief technical overview of the biochemistry of vision. His brief overview required three-and-a-half pages of text. Let me quote just the first paragraph of his summary (pp. 18–19) to give you a sense of just how impossibly complex this is.

"When light first strikes the retina a photon interacts with a molecule called 11-cis-retinal, which rearranges within picoseconds to trans-retinal.

Irreducible Complexity

Professor Behe knows many, many things that Charles Darwin never had any chance to know. Behe knows that at a molecular level, living things are truly amazing. No doubt, Darwin probably thought about the stuff of living things as "protoplasm"—a vague nineteenth-century reference to the living substance of cells. But Behe knows that each and every cell of the human body is an amazing micro-factory—chock-full of organelles performing a wide variety of complex functions. The precision and efficiency of these micromachines is truly astonishing.[94]

As we all know, most machines can't function with a few parts missing—and often with only a single part missing. Professor Behe's term for this is "irreducible complexity." He illustrates this by reference to a machine everyone understands, an ordinary mousetrap. The mousetrap has five basic parts: a platform, a holding bar, a spring, a catch, and a hammer (plus a few staples to hold things together). Remove even one of the basic parts and the trap doesn't work at all.

(A picosecond is about the time it takes light to travel the breath of a single human hair.) The change in the shape of the retinal molecule forces a change in the shape of the protein, rhodopsin, to which the retinal is tightly bound. The protein's metamorphosis alters its behavior. Now called metarhodopsin II, the protein sticks to another protein, called transducin. Before bumping into metarhodopsin II, transducin had tightly bound a small molecule called GDP. But when transducin interacts with metarhodopsin II, the GDP falls off, and a molecule called GTP binds to transducin. (GTP is closely related to, but critically different from, GDP.)"

Behe's description of the biochemistry of vision continues for four more, somewhat longer, similar paragraphs.

[94] To give just one tiny example among many, Wells has observed that "cells with nuclei contain microscopic fibers called 'microtubules.' Molecular motors travel along the microtubules, transporting various cargoes throughout the cell." (Wells, *Zombie Science*, 91).

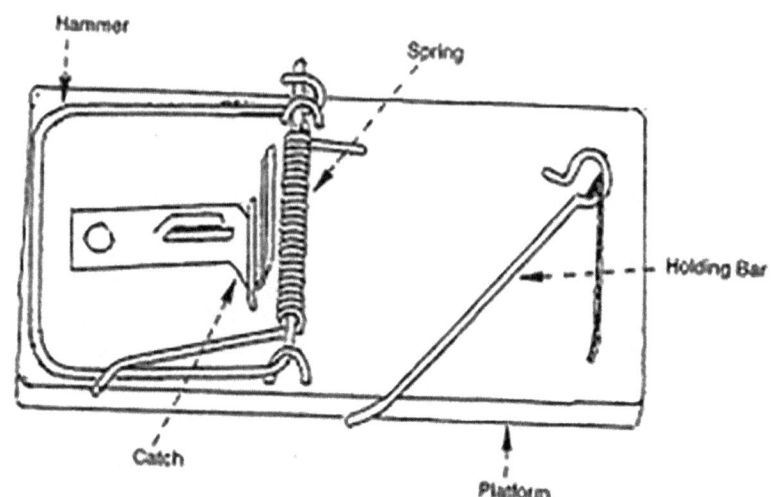

Figure 8. Behe's mousetrap, from Darwin's Black Box, page 43. (Copyright Michael J. Behe, 1996. Used by permission.)

Professor Behe defines irreducible complexity this way:

> By irreducibly complex I mean a single system composed of several well-matched, interacting parts that contribute to the basic function, wherein the removal of any one of the parts causes the system to effectively cease functioning. An irreducibly complex system cannot be produced directly (that is, by continuously improving the initial function, which continues to work by the same mechanism) by slight, successive modifications of a precursor system, because any precursor to an irreducibly complex system that is missing a part is by definition nonfunctional.[95]

Behe knows that the idea of such complex, integrated biological systems arising suddenly rather than gradually "is almost universally

---

[95] Behe, *Darwin's Black Box,* 39.

conceded [to] be irreconcilable with the gradualism Darwin envisioned."[96] Even the most radical Darwin defender, Richard Dawkins, sees the problem.

> Evolution...must be gradual when it is being used to explain the coming into existence of complicated, apparently designed objects, like eyes. For if it is not gradual in these cases, it ceases to have any explanatory power at all. Without gradualness in these cases, we are back to miracle.[97]

And Dawkins also knows that the issue of irreducible complexity is a make-or-break issue for Darwin's theory. In 1986, he wrote:

> One hundred and twenty-five years on, we know a lot more about animals and plants than Darwin did, and still not a single case is known to me of a complex organ that could not have been formed by numerous successive slight modifications. I do not believe that such a case will ever be found. If it is—it'll have to be a really complex organ—*I shall cease to believe in Darwinism.*[98]

As we shall see, the mechanism by which evolutionists continue to dismiss the huge problem of irreducible complexity is simply by refusing to recognize facts that are staring them in the face.

Eyes Tight Shut

From all appearances, even as one of the world's most famous advocates of Darwin's big idea, Richard Dawkins simply doesn't understand the idea of irreducible complexity. One of his best-selling

---

96 Ibid., 40.
97 R. Dawkins, *River Out of Eden* (New York: Basic Books, 1995), 83.
98 Dawkins, *The Blind Watchmaker*, 91. (Emphasis added.)

books promoting Darwin's big idea was called *The Blind Watchmaker*.[99] In that book, Dawkins took great issue with critics of Darwin who argued that complex biological systems would have no practical value unless they had come into being all at once. This was another way of saying that biological systems are irreducibly complex because a biological machine missing a few parts would not work at all.

For example, one such critic had said,

> It will be seen that each one of this sequence of conditions is essential for the success of the whole. Yet each by itself is useless. This whole *opus perfectum* must have been achieved simultaneously. The odds against the random occurrence of such a series of coincidences are, as we have already stated, astronomical.[100]

Yet Dawkins disagreed emphatically.

> It isn't true that each part is essential for the success of the whole. A simple, rudimentary, half-cocked eye…is better than none at all. Without an eye you are totally blind. With half an eye you may at least be able to detect the general direction of a predator's movement, even if you can't focus a clear image.[101]

"Half an eye?" Really? What exactly can be seen by "half an eye?" In the context of Michael Behe's model, perhaps it is easier to ask how many mice will you catch with half a mouse trap? Obviously, the answer is none. Nor will an eye missing half its parts allow you to see anything at all, including anything about a predator's movement.

---

[99] Cited first at footnote 4.

[100] Ibid., 41, quoting Bishop Montefiore, *The Probability of God* (London: SCM Press 1985), which in turn is quoting C. E. Raven on cuckoos.

[101] Ibid.

But surely Dawkins must have understood this. Surely we must be missing something about what he was trying to say, yes?

No, we are not missing anything. Because a bit further on in his very famous book, Dawkins doubles down on his argument concerning the value of a part of an eye.

He does this by taking issue with another well-known evolutionist, Stephen Jay Gould. Gould had asked the question "What good is 5 percent of an eye?" Gould had assumed that 5 percent of an eye is good for nothing and had speculated that if part of a complex eye had been preserved by natural selection, it must have been used by the creature possessing it for some purpose other than sight.

Again, Dawkins disagreed with his colleague from Harvard.

> An ancient animal with 5 per cent of an eye might indeed have used it for something other than sight, but it seems to me at least as likely that it used it for 5 per cent vision…Vision that is 5 per cent as good as yours or mine is very much worth having in comparison with no vision at all. So is 1 per cent vision better than total blindness. And 6 per cent is better than 5, 7 per cent better than 6, and so on up the gradual, continuous series.[102]

So Dawkins appears to actually believe that 5 percent of an eye actually produces some vision. Remember, he's not talking about 100 percent of some primitive eye but 5 percent of a complex eye. Dawkins is making the logical mistake of thinking that a certain percentage of a machine will accomplish the same percentage of the purpose for which the whole machine exists. But purpose and practical reality are two very different things.

Dawkins' contention reflects some extremely sloppy thinking. Behe is right and Dawkins is wrong. The plain fact is that even a machine as simple as a mousetrap *cannot work at all* unless all the parts of the machine are in place. So 5 percent of any machine should

---

[102] Ibid., 81.

not be expected to produce 5 percent of anything at all, much less something as amazing as even a tiny bit of sight.

What can explain the willingness of a man like Dawkins who holds a doctorate from Oxford University to claim that half an eye—or even 5 percent of an eye—can actually see something? I think the only explanation is one applicable to many, many advocates of Darwin's big idea, including Darwin himself. It is the combination of an overactive imagination coupled with an overpowering desire to support an idea that, in fact, is without support. It is storytelling rather than science.

As you read the following summaries of a few irreducibly complex biological systems, ask yourself whether 5 percent of the system being described would produce a result that is useful in any way. If you think about it at a molecular level, just 5 percent of virtually every biological tissue, organ, or system on earth could not possibly function normally. Moreover, just 5 percent of any of the cells of which such things are constructed would also be completely unable to function because each of those cells is itself irreducibly complex.

The Blood-Clotting System

Behe's book *Darwin's Black Box* explains several examples of irreducibly complex biological systems at great length, one of which will be summarized here.

There's little doubt that Charles Darwin was cut and bled from time to time during his life. He knew his bleeding would stop, but he had no idea why. The details of the human blood-clotting system were not understood for more than one hundred years after Darwin died.

Think about it. If you poke a hole in the side of a container of water, does the water stop running out before the container is empty, especially if the water is under pressure? So why does a person not lose all his blood when he gets cut? The answer is far from simple. And the answer is even more complicated because the blood-clotting system has to do opposite things at exactly the right time for each. When blood is flowing out of a cut, the system has to clot the blood and stop the flow. If this clotting system fails, the person dies.

But when blood is flowing normally through *uncut* vessels to sustain living tissue, the system must *not* stop the flow with a blood clot. If such a clot occurs at the wrong time, the person also dies.

In fact, the answer to why blood clots (or not) is so vastly complicated that it takes more than ten pages of text for Professor Behe to provide even a simple overview of how the blood-clotting system works. Since a picture is worth a thousand words, I'll present Behe's summary by showing you his "flowchart" of the blood-clotting system, starting from the "Wound Surface" and ending in "Fibrin (hard clot)."

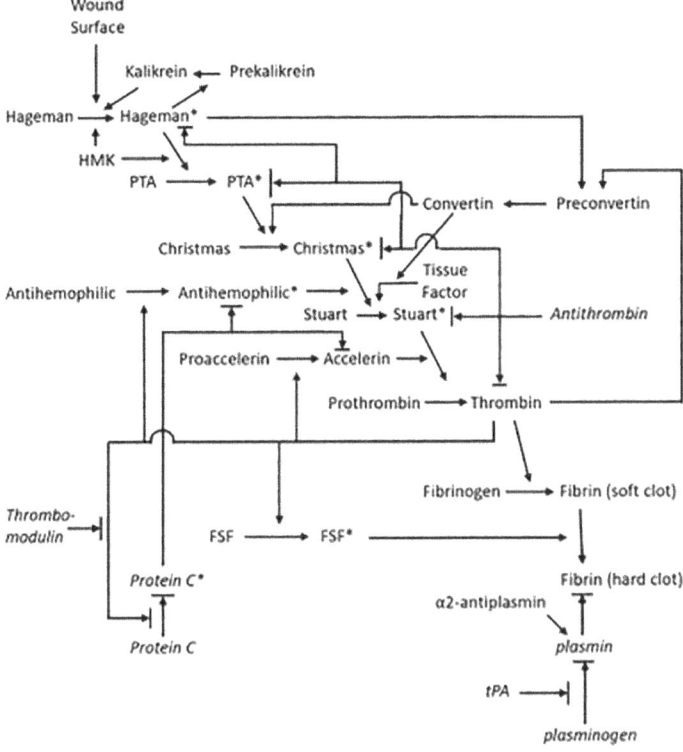

Figure 9. Blood clotting cascade very similar to the diagram presented in Darwin's Black Box, page 82. Arrows ending in a bar indicate proteins acting to prevent, localize or remove blood clots. (Copyright Michael J. Behe, 1996. Used by permission.)

Yes, this amazing chart seems impossible for the average person to comprehend. What you are looking at are the names of a series of proteins, enzymes, and proenzymes that communicate with one another in an extremely complicated way. It is referred to as the blood-clotting "cascade" where one component of the system activates the next, which then activates a third component, which then activates a fourth, and so on. The chart also includes the many factors responsible for *stopping* the clotting process at just the right time so the creature's entire circulatory system does not clot.

Overall, this blood-clotting cascade involves multiple factors that enable clotting exactly when it is needed and multiple factors that disable clotting when it is no longer needed. The life-and-death question is this: "Are these factors in exactly the right balance or not?" If yes, the animal lives. If no, the animal is in great trouble or is dead.

Using assigned names for each of the pro-clotting and anti-clotting factors, Professor Behe depicts this crucial balance as follows:

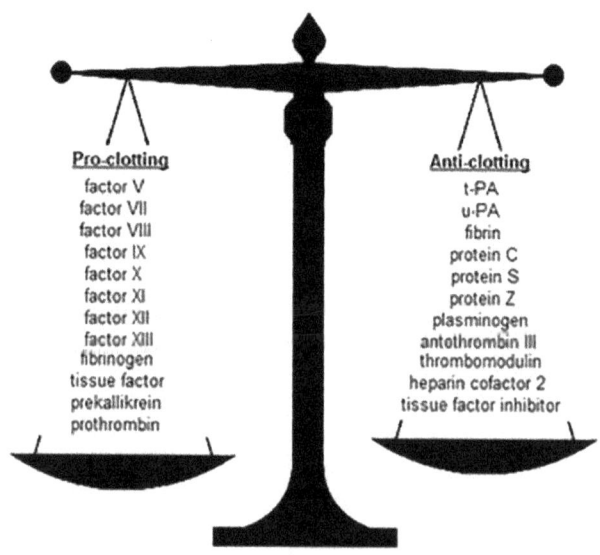

Figure 10. Clotting factor balance. Information taken from Behe, Darwin Devolves, page 301.

Does all this look a bit more complicated than a mousetrap? How would our subject animal be doing if all of its anti-clotting factors had evolved but only half of its pro-clotting factors? Or the opposite? Or just 5 percent of each? Is this a biological system that could have arisen by a long slow series of slight modifications? The undeniable answer is no, not if our subject wants to be able to survive and reproduce.

Amazingly, no scientist has ever even tried to offer a serious fact-based estimation of how the blood-clotting cascade might have arisen slowly by mutation and natural selection. In 1993, a scientist named Richard Doolittle wrote about the state of scientific knowledge concerning the evolution of blood clotting in a journal called *Thrombosis and Haemostasis*—a journal dedicated entirely to the topic of blood clotting and read by the handful of people who know more about blood clotting than anyone else on earth. Professor Doolittle was himself at the time probably the world's leading expert on the supposed evolution of blood clotting.

Professor Doolittle's article can barely be described; it can only be quoted. So here is an extended quote, with Professor Behe's slight simplification of technical terms:

> Blood clotting is a delicately balanced phenomenon involving proteases, antiproteases, and protease substrates. Generally speaking, each forward action engenders some backward-inclined response. Various metaphors can be applied to its step-by-step evolution: action-reaction, point and counterpoint, or good news and bad news. My favorite, however, is yin and yang.
>
> In ancient Chinese cosmology, all that comes to be is the result of combining the opposite principles of yin and yang. Yang is the masculine principle and embodies activity, height, heat, light and dryness. Yin, the feminine counterpoint, personifies passivity, depth, cold, darkness and wet-

ness. Their marriage yields the true essence of all things. Keeping in mind that it's only a metaphor, consider the following yin and yang scenario for the evolution of vertebrate clotting. I have arbitrarily designated the enzymes or proenzymes as the yang, and the nonenzymes as the yin.

Yin: Tissue Factor (TF) appears as the result of the duplication of a gene for [another protein] that binds EGF domains. The new gene product only comes into contact with the blood or hemolymph after tissue damage.

Yang: Prothrombin appears in an ancient guise with EGF domain(s) attached, the result of a... protease gene duplication and...shuffling. The EGF domain serves as a site for attachment to and activation by the exposed TF.

Yin: A thrombin-receptor is fashioned by virtue of the duplication of a gene for a [protein region that will stick in a cell membrane]. Cleavage by the TF-activated prothrombin effects cell contractility or clumping.

Yin again: Fibrogen is born, a bastard protein derived from a thrombin-sensitive [elongated] father and a [protein with a compact structure for a] mother.

[And the article continues on with a dozen more yin and yang contributions to the system.][103]

---

[103] Behe, *Darwin's Black Box*, 91–93, quoting R.F. Doolittle (1993) "The Evolution of Vertebrate Blood Coagulation: A Case of Yin and Yang," *Thrombosis and Haemostasis*, 70, 24–28.

Huh? A yin-and-yang metaphor to explain the origin of a system that is absolutely essential to human life? Moreover, an explanation that names no actual causes but simply speculates that tissue factor "appears" or that fibrogen "is born" or that something is the result of "shuffling"? Shuffling what? Shuffling for how long? Again, this is storytelling, not science.

In Doolittle's imaginations, the first sign of clotting ("clumping") does not even appear until the third step in the process. But according to Darwin's big idea, natural selection would have eliminated those first two steps if they had not yet resulted in an adaptation useful for life and reproduction.

Professor Behe explains the actual bottom line of this curious "scientific" article:

> Doolittle's scenario implicitly acknowledges that the clotting cascade is irreducibly complex, but it tries to paper over the dilemma with a hail of metaphorical references to yin and yang. The bottom line is that [in order for the system to work] clusters of proteins have to be inserted *all at once* into the cascade.
>
> [T]he article does not explain...how clotting might have originated and subsequently evolved: instead, it just tells a story. The fact is, *no one on earth has the vaguest idea how the coagulation cascade came to be.*[104]

The Whale's Testicular Cooling System

As noted above, biologists seem absolutely convinced that whales evolved from some earlier land-based mammal (see chapter 3, figures 6 and 7.) Even Darwin himself speculated that a creature something like a black bear might have become more and more aquatic over time until it

---

[104] Behe, *Darwin's Black Box*, 96–97. (Emphasis original.)

became "as monstrous as a whale."[105] But neither Darwin nor any of his followers has ever found fossil evidence providing anything like a series of transitional forms showing how some four-legged mammal living on land actually became a multi-finned mammal living in the ocean.

It may be easy to forget that whales are mammals that breathe air into lungs through a blowhole in the top of their heads. There are several other things that all mammals do that are not so easy to do underwater—such as whales nursing their young with milk. The mother whale actually "forcefully squirts milk into the [mouth of the] calf"—milk that is "three to four times as concentrated as the milk of cows or goats; it has the consistency of condensed milk or liquid yogurt. The calf thereby receives much more nourishment in a much shorter time."[106]

So how exactly could it be that a land mammal could have any of its young survive if it began to live underwater prior to the development of the thick milk power squirt system? And why would such a complex system even begin to evolve if the creature was still spending enough time on land to nurse its young in the way common to land mammals? You see the problem.

Yet perhaps the most amazing anatomical difference between any land-based mammal ancestor and the whale is the unique system by which male whales cool their testicles to allow sperm production. In mammals, "sperm production requires a temperature several degrees below normal body temperature."[107] For land-based mammals, the solution is easy: the testicles are housed in a scrotum suspended in cooler air below the body. In male whales, however, the testicles are within the body surrounded by heat-generating muscles.[108] So how is the cooling accomplished for whales? In a way that is most definitely irreducibly complex.

> The cooling is accomplished with a counter-current heat exchanger. Blood that has been cooled

---

[105] Darwin, *Origin*, 215.
[106] Wells, *Zombie Science: More Icons of Evolution*, 109.
[107] Ibid., 107.
[108] Ibid.

in the dorsal fin and flukes is carried to a region near the testicles, where it flows through a network of veins that pass between arteries carrying warm blood in the opposite direction. The arterial blood is thereby cooled before it reaches the testicles.[109]

Did you catch that? Not just a single set of vessels that carry blood to the testicles from the cooler dorsal fin, *but a double set of vessels in which blood flows in opposite directions so that cooler blood can reduce the temperature of warmer blood!*
A very simplified diagram of this system is the following:[110]

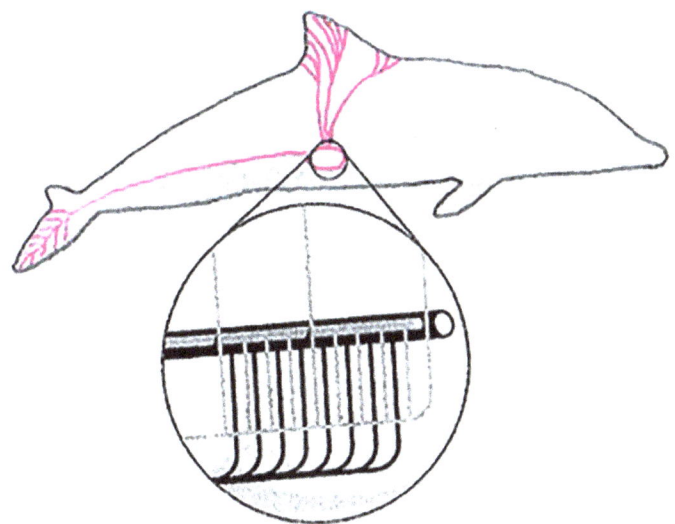

Figure 11. Simplified diagram of internal testicle cooling system. Red lines show cooler blood from dorsal fin and tail flukes. Inset shows warm arterial blood (black) being cooled by adjacent cooler veins (gray). (Copyright Discovery Institute, 2017. Used with permission.)

---

[109] Ibid., citing Sentiel A. Rommel, D. Ann Pabst, William A. McLellan, James G. Mead, and Charles W. Potter, "Anatomical evidence for a countercurrent heat exchanger associated with dolphin testes," *Anatomical Record* 232 (1992), 150–156.
[110] Ibid., 108.

Now think for a moment about how this amazing counter-current cooling system could have evolved. Darwin's big idea would suppose that the external testicles of the whale's land-based ancestor somehow slowly moved more and more inside the body of the water-based whale over a vast period of time. But when in that long process, would this cooling system have come to be?

> If this engineering arrangement were due to evolution, the relocation of the [whale's] testicles to the inside could not have preceded the counter-current heat exchange system. Otherwise, the whale would have been sterile, an evolutionary dead end. Yet there is no adaptive advantage to developing a counter-current heat exchange system around the testicles unless they are inside the body. One would not come before the other, yet the probability that both would evolve simultaneously is effectively zero.[111]

So again, you see the problem. And the problem goes far beyond the difficulty of imagining how or why such a complex system could have evolved by numerous tiny steps controlled by forces that had no particular outcome in mind—indeed, by forces that are literally mindless. The problem here is one of logical impossibility. In order for the creatures to survive, these difficult anatomical changes could not have occurred before a land mammal left dry ground but had to be fully functional before a water-based whale took to the sea. But even according to Darwin, there is no possibility that all this could have occurred instantly in fully functional form. Nor is there reason to think the changes would have even begun before the land mammal was almost fully aquatic. So for the evolutionist, this ingenious cooling system is a logical impossibility.

---

[111] Ibid.

Sexual Reproduction

In the preface of this book, I told you that I grew up on a farm that produced feeder lambs for meat. One of my jobs at this Northern Wisconsin location (where winter temperatures actually got down to *minus fifty degrees* on the thermometer!) was to get up in the middle of the night to see if any lambs had been born so I could towel them off and be sure they were dry and nursing instead of wet and freezing. Consequently, I ended up helping in the delivery of lots of lambs, the usually joyful end product of a several month's long series of events. So even as a young boy, sexual reproduction was something very familiar to me.

Amazingly, evolutionists have yet to figure out how sexual reproduction could have possibly evolved.

Many of the simplest creatures on earth reproduce asexually by a variety of means. Certain bacteria reproduce by a process called binary fission, where a single parent divides into two genetically identical daughters. A tiny multicelled creature called a hydra reproduces by rapid cell division at one particular site of the creature producing a small projection called a bud, which later breaks off and develops into a genetically identical daughter hydra. Fungi and certain algae reproduce by producing spores which, after dispersal, develop into new organisms genetically identical to the original.

Sexual reproduction differs fundamentally from these methods in that it involves two parents, a male and a female, each of which contributes only half of the genetic information possessed by the offspring. With this method, the new plant or animal is never genetically identical to either parent.

Evolutionists are quick to say that sexual reproduction must have begun for the purpose of creating beneficial genetic diversity in a population. And while that may explain *why* it began, Darwinists have no idea at all about *how* it began.

Some of the more honest observers—including some proponents of evolution—admit as much. For example:

> Despite some ingenious suggestions by orthodox Darwinians, there is no convincing Darwinian history for the emergence of sexual reproduction.[112]

> Sex is a puzzle that has not yet been solved; no one knows why it exists.[113]

> [T]he existence of sexual reproduction poses a big theoretical puzzle for Darwinians.[114]

> Why sex? At first blush, its disadvantages seem to outweigh its benefits. After all, a parent that reproduces sexually gives only one-half its genes to its offspring, whereas an organism that reproduces by dividing passes on all its genes. Sex also takes much longer and requires more energy than simple division. Why did a process so blatantly unprofitable to its earliest practitioners become so widespread?[115]

> The overriding question is when (and then how) sexual reproduction has evolved. Despite decades of speculation, we do not know.[116]

---

[112] Graham Bell, *The Masterpiece of Nature: The Evolution and Genetics of Sexuality* (Berkeley, CA: University of California Press), 54.

[113] Mark Ridley, *The Cooperative Gene* (New York: The Free Press, 2001), 111.

[114] Dawkins, *The Blind Watchmaker*, 268.

[115] Julie Schecter, How Did Sex Come About?, *Bioscience*, 34:680, December, 1984.

[116] John Maddox, *What Remains to be Discovered* (New York: The Free Press, 1998), 252.

But even if we suppose that genetic diversity explains the "why" of sexual reproduction, what in the world is the answer to the question of "how"?

Remember Darwin's own admission about biological complexity:

> If it could be demonstrated that any complex organ existed, which could not possibly have been formed by numerous, successive, slight modifications, my theory would absolutely break down.[117]

Darwin's use of the phrase "complex organ" was an extreme understatement of the problem. He knew very well that many functions essential to life depend not on a single organ but on a system involving multiple complex organs. Moreover, such essential organ systems always require other biochemical factors such as hormones—signaling molecules—to tell them what to do and when to do it. So the fact is that Darwin's big idea must be able to explain not only the creation of every single organ but the creation of entire vastly complex biological systems. If Darwin had been completely well-informed and honest, he should have not been talking about individual organs. He should have admitted that if it could be demonstrated that any *complex biological system* existed which could not possibly have been formed by numerous, successive, slight modifications, then his theory would absolutely break down.

Sexual reproduction is an excellent example of just such a biological system which cannot have arisen as Darwin supposed. To help you reach this important conclusion, I will note just a few relatively simple things about this life-or-death system. I will talk in terms of mammals since we are mammals and our reproductive systems are very familiar to us. But please don't forget that these few notes are a *vast understatement* of the complexity of the actual system—which

---

[117] Darwin, *Origin*, 219.

would take hundreds (if not thousands) of pages to describe in great detail.[118]

Phase 1: Successful Gamete Generation

In the male and the female, viable gamete production requires overall development of the adult through puberty to sexual maturity—which is itself a complex, multifactor process. In the male, a fully functional testicle must have properly descended into a scrotum for proper temperature control. The functional testicle contains several different types of cells with very specific purposes concerning the production, maturation and protection of viable sperm, including the production of testosterone. In the female, a fully functional ovary contains at birth all the egg cells she will ever possess. Upon arriving at sexual maturity, her brain signals the pituitary gland to begin producing hormones which cause the ovary to produce still other hormones that result in the maturation of one or more eggs, month by month.

Phase 2: Successful Gamete Combination

In the male, viable sperm must move from the epididymis where they are stored through the vas deferens to be mixed with secretions from the seminal vesicles, the prostate gland, and the bulbourethral gland which altogether comprise the seminal fluid. Upon orgasm, this fluid is injected into the body of the female only through an erect penis. The process of erection and ejaculation is itself an extremely complex and delicate combination of neural, hormonal, and anatomical factors. In the female,

---

[118] My summary of sexual reproduction is taken from a variety of sources, none of which I believe to be the least bit controversial. Nevertheless, my dear daughter, Abigail E. Kirk, has double-checked the accuracy of this description as she is a doctor of nursing practice graduate of the University of Minnesota, an advanced practice registered nurse, and a certified nurse-midwife. Any errors in the text are mine, not hers.

the mature egg is released by the ovary and caught by the fimbriae, which are lined with hundreds of thousands of microscopic fingers gently brushing the egg onward in a mechanical, wavelike motion. The egg then descends through the fallopian tube to the uterus. This process of egg maturation and release is controlled by a complicated system of hormonal secretions. Somewhere in this process, the egg is found and penetrated by a sperm cell which has been deposited in the vagina and passed through an unsealed cervix into the uterus (or perhaps into a fallopian tube).

Phase 3: Successful Gestation of the Zygote/Fetus

The fertilized egg, now known as the zygote, cannot develop properly unless it implants in the endometrial lining of the uterine wall. Many people are familiar with the concept of "implantation," but what is actually happening over a relatively short period of time is so vastly complex that it is almost beyond words. Here is just a hint:

- During its first couple days in the uterus, the zygote somehow first develops something known as the "yolk sac" which later becomes part of the developing gut of the fetus. Before the development of the placenta, the yolk sac provides both nutrition and a gas exchange mechanism between the mother and the developing embryo.
- The zygote then somehow develops an umbilical cord, one end of which begins at its belly and the other end of which attaches to the wall of the uterus where the placenta develops. Using the umbilical cord to connect the mother's blood supply to the fetus, the placenta provides a mechanism for fetal nutrition, temperature regulation, and gas exchange.
- The zygote is also quickly enclosed by a double sac of protection. The inner sac is known as the amnion which,

within weeks, begins to fill with amniotic fluid, providing additional protection for the fetus until delivery. The outer sac is called the chorion. The chorion is covered with minute, wormlike processes called chorionic villi between the chorion and the placenta, which increase fetal access to maternal blood.

Phase 4: Successful Delivery of the Fetus

Successful delivery of the fetus begins with the fact that the female pelvis has a different shape than the male pelvis to make it easier for the fetus to emerge. Primarily, the female pelvis is wider, allowing for a larger birth canal. A fully functional uterus performs not only as the location for the fetus to develop fully but also as a major part of the muscles used to deliver the fetus. Delivery is accomplished by maternal muscles, both voluntarily and involuntarily, pushing the fetus through the dilated cervix and out of the mother's body. What is actually happening at a biochemical level is a cascade of several different hormones which both prepare certain muscles to contract, later actually cause them to contract, and at the same time prepare the tissues of the cervix for delivery. The oxytocin receptors must be kept quiet during pregnancy so they are less likely to respond to signals which could stimulate contractions. Later in pregnancy, these receptors are actually multiplied in order to produce the overwhelming response to oxytocin required for the regular progressive labor contractions as well as the life-preserving contractions immediately after delivery, which prevent post-partum hemorrhage.

Phase 5: Successful Nursing of the Infant Mammal

Infant mammals would die quickly if not able to nurse milk from their mother. This requires the mother to have a functioning breast that produces an adequate amount of milk.

The mammary gland is a relatively complex organ that literally converts blood into milk by means of a series of glands drained by a series of ducts which deliver it through a nipple when nursed. Milk must be produced at just the right time, a process controlled by specialized hormones. The first milk produced, known as colostrum, contains many special ingredients to give the newborn a strong start in life. These include antibodies to ward off disease and infection, immune and growth factors, and other bioactives that help activate the immune system, jump-start gut function, and promote a healthy gut during the challenge of digestion beginning. The infant must also be willing to nurse—itself an amazing instinct of mysterious origin.

Exact Hormonal Control of All Five Phases

The brief notes of five phases of sexual reproduction summarized above have focused on the organs and structures needed for sexual reproduction, but those structures are not nearly enough. Biochemistry is also essential. Each of the phases described are regulated by a complicated series of hormones. You may have heard of some of these, but many are almost never discussed: estrogen, progesterone, testosterone, estriol, cortisol, prostaglandins, relaxin, oxytocin, gonadotropin-releasing hormone, follicle-stimulating hormone, luteinizing hormone, chorionic gonadotropin, and the list goes on and on. Each of these requires still other structures or factors for their manufacture, release, and exact control of their precise quantity and timing.

•

Think for a moment about the magnitude of the problem of successful sexual reproduction. It is truly enormous. Since Darwin's big idea is based on processes that are truly random, mindless, and unplanned, what are the chances that all of the things that would have to develop in a fully functional male would come to completion in the lifetime of one fully functional male or of one fully functional female? And then there's the blockbuster question: "What are the chances that those two unique individuals, a fully functional male and female, would be completed by mindless evolution in the same place and at almost exactly the same time so that they could actually mate and produce offspring to carry on the vastly complex invention of sexual reproduction?" And what are the chances that this same impossible coincidence would have happened for each of the millions of sexually reproducing species on earth?

A simple mousetrap truly is irreducibly complex. But sexual reproduction by mindless Darwinian mechanisms truly is an absolute impossibility.

No Explanation of Any Kind for Any Complex Organ or System

Blood clotting, the male whale sex organs, and sexual reproduction are just three examples from the uncountable number of complex organs and systems embodied in the millions of species of plants and animals on earth that have *no plausible explanation* based on Darwin's big idea.

And it is important to note that this complete absence of plausible explanations based on mutation and natural selection is true not only for a great many particular biological organs and systems, but it is also true for some of the most basic and common of all biological building blocks—things like a living cell and DNA.

In two books written almost thirty years apart, Michael Denton described the problem in the context of the individual cell:

> In *Evolution: A Theory in Crisis* I wrote: "Between the living cell and the most highly ordered non-biological system...there is a chasm as vast and absolute as it is possible to conceive." Thirty years on, the situation is entirely unchanged. Despite a vast increase in knowledge of supra-molecular chemistry and of cell and molecular biology... no one has provided even the vaguest outlines of a feasible scenario, let alone a convincing

one. A yawning gap still persists—empirical and theoretical.[119]

The situation is exactly the same concerning DNA itself.

[N]o plausible scenario for the evolution of the modern DNA-protein genetic code, via gradual functional continuums of increasingly more complex cellular forms, has ever been developed.

At the heart of the problem is a dreary vicious circle: what would be the selective force behind the evolution of the extremely complex translation system before there were functional proteins? And, of course, there could be no [functional] proteins without a sufficiently effective translation system.[120]

And when Denton says there is "no plausible scenario," he means literally *no scenario*.

Perhaps the most valuable single contribution of Professor Behe's book *Darwin's Black Box* was made by chapter 8 of that book titled "Publish or Perish." In that chapter, Behe reported the results of his truly exhaustive search of the literature, looking for any scientific explanation—or even hypothesis—at the molecular level of how any complex organ or biological system could have actually arisen by mutation and natural selection. Behe's search did not just look at the titles of journal articles or the indexes of books, but when he found a promising entry, he looked further to see if the author was offering an actual explanation.

---

[119] Denton, *Evolution: Still a Theory in Crisis in Crisis*, 121.

[120] Ibid., 124, quoting Eugene V. Koonin and Artem S. Novozhilov, "Origin and Evolution of the Genetic Code: The Universal Enigma," *IUBMB Life* 61, no. 2 (February 2009), 108.

For example, Behe reports finding a book index reference to "evolution, adaptation of sperm whale."

> When we flip to the indicated page, we learn that sperm whales have several tons of oil in their heads which becomes more dense at colder temperatures. This allows the whale to match the density of the water at the great depths where it often dives and so swim more easily. After describing the whale the textbook remarks, "Thus we see in the sperm whale a remarkable anatomical and biochemical adaptation, perfected by evolution." But that single line is all that's said! The whale is stamped "perfected by evolution," and everybody goes home. The authors make no attempt to explain how the sperm whale came to have the structure it has.[121]

Behe's literature search started with the *Journal of Molecular Evolution*, established in 1971 and "devoted exclusively to research aimed at explaining how life at the molecular level came to be."[122] He found nothing.

> In fact, *none* of the papers published in JME over the entire course of its life as a journal has ever proposed a detailed model by which a complex biological system might have been produced in a gradual, step-by-step Darwinian fashion.[123]

---

[121] Behe, *Darwin's Black Box*, 181.
[122] Ibid., 165.
[123] Ibid., 176. (Emphasis original.)

From there, he reviewed about twenty thousand life sciences papers published during the years 1984–1994 as the *Proceedings of the National Academy of Sciences*. The result was the same.

> No papers were published in PNAS that proposed detailed routes by which complex biochemical structures might have developed. Surveys of other biochemistry journals show the same result…[124]

From there, Behe turned to the world of books. Again, the result was the same—a finding that did not surprise Behe because "[t]he absence of papers on the evolution of biochemical structures in the journals just about kills any chance of there being a book published on the matter."[125]

Overall, Behe's exhaustive search led him to the conclusion that there is *no scientific authority at all* for Darwin's big idea at the molecular level of life.

> *There is no publication in the scientific litera-ture*—in prestigious journals, specialty journals, or books—*that describes how molecular evolution of any real, complex, biochemical system either did occur or even might have occurred…*If a theory claims to be able to explain some phenomenon but does not generate even an attempt at an explanation, then it should be banished.[126]

Professor Behe's most recent book reported that *as of 2019, there still was no publication of any kind that even attempts to explain how any complex biological system could have arisen by a Darwinian mecha-*

---

[124] Ibid., 178.
[125] Ibid.
[126] Ibid., 185–86. (Emphasis added.)

nism. Referring back to what he said in 1996 in *Darwin's Black Box,* Behe writes,

> More than two decades later—despite the uproar caused by the book, despite much bluster and chest thumping in the media—the situation is unchanged. *The literature remains totally devoid of explanations,* and Darwinists remain incongruously smug.[127]

In *Darwin's Black Box,* Behe argued concerning the absence of relevant articles "the theory of Darwinian molecular evolution has not published, and so it should perish."[128] This conclusion might sound extreme, but it is easy to argue that Darwin himself actually might have agreed. Remember, Darwin said, "If it could be demonstrated that any complex organ existed, which could not possibly have been formed by numerous, successive, slight modifications, *my theory would absolutely break down.*"[129]

For this reason, among many others, reasonable people should conclude that Darwin's big idea has indeed broken down.

---

[127] M.J. Behe, *Darwin Devolves: The New Science About DNA That Challenges Evolution* (New York, NY: HarperCollins, 2019), 216. (Emphasis added.)

[128] Behe, *Darwin's Black Box,* 186.

[129] Darwin, *Origin,* 219. (Emphasis added.)

CHAPTER 5

MICROEVOLUTION HAPPENS BY MEANS THAT CANNOT PRODUCE MACROEVOLUTION

Accidents happen. The results are almost never good. But every once in a great while, an accident can produce something good. Apparently the multibillion-dollar dry cleaning industry was born when the careless maid of a textile worker spilled his kerosene lamp onto a dirty tablecloth and he saw that, when it had dried, certain stains had been removed. Similarly, the antibiotic later developed and named penicillin was first noticed when some mold unintentionally made its way into a dish of bacteria and killed the bacteria it touched.

But an accident is still an accident. People commonly do their best to avoid accidents because we know that multibillion-dollar industries or life-saving medicines are almost never the result.

Mutations are accidents—genetic copying errors. Yet Darwin's big idea presumes that these accidents somehow have managed to produce billions of biological machines so complicated as to be beyond even our ability to understand, much less recreate.

Despite what common sense tells us about the value of accidents, proponents of Darwin's big idea have long repeated the vague assertion that mutations have been the source of the "raw material" required for macroevolution to work. As early as 1937, a major work synthesizing Darwin's big idea with the science of genetics declared, "Mutations and chromosomal changes…constantly and unremittingly supply the raw materials for evolution."[130]

---

[130] Theodosius Dobzhansky, *Genetics and the Origin of Species* (New York: Columbia University Press, 1937), 13.

But do they? One imaginative metaphor puts the problem this way: "[O]ne can sometimes "fix" a sputtering radio by hitting its case if the rough motion happens to reset a loose wire or open a short circuit. But no one would expect to build a better radio, much less a television set, by accumulating such changes."[131]

So do mutations actually supply the raw materials for Darwin's big idea? The most recent science actually investigates the problem directly for the first time, and these latest studies teach us that the answer to our question is no.

To study mutations efficiently, one needs to be working with a species that produces multiple generations in a very short period of time. Accordingly, fruit flies have been studied extensively since the early 1900s because they can be grown from eggs to adults in just seven days and the females lay hundreds of eggs during a life span that lasts only a few weeks.

During the last one hundred-plus years, many, many millions of generations of fruit flies have been bred and studied in laboratories around the world. More importantly, the number of mutations occurring in those many generations has been increased by exposing the flies to artificial factors such as radiation. Many thousands of such mutations have been observed. A few have no effect on the fly, but the great majority of them have been harmful, if not fatal.[132] The overall lesson from this research has been very clear.

> All of the evidence points to one conclusion. No matter what we do to the DNA of a fruit fly embryo, there are only three possible outcomes: a normal fruit fly, a defective fruit fly, or a dead fruit fly. Not even a horse fly, much less a horse.[133]

---

[131] Johnson, *The Wedge of Truth*, 47.
[132] Wells, *Icons*, 178.
[133] Wells, *Zombie Science*, 94.

Yet once in a very great while, a mutation does appear to confer an advantage on the creature that experiences the mutation. The most recent genetic studies of these populations now show that even these "advantages" are the result of genetic information being *damaged or destroyed*, not added. If the mutations driving evolution always subtract from genetic information rather than add to it, then this subtractive process cannot possibly produce the "raw material" needed for complex new organs, body plans, or biological systems.

In other words, a central premise thought to support the extrapolation from micro to macroevolution is false.

Michael Behe's 2019 book *Darwin Devolves: The New Science About DNA That Challenges Evolution* makes this point at great length.[134] The central thesis of the book relies on the most modern genetic technology that has enabled researchers to examine the entire genomes of populations in which microevolution has been observed, starting with Darwin's Galapagos finches.

Darwin's Finches Don't Prove that His Big Idea
Happened but Help Show That It Can't

Charles Darwin visited the Galapagos Islands in 1835 and collected a variety of specimens, including several varieties of finches. More than one hundred years later, a British ornithologist named David Lack wrote a book called *Darwin's Finches* in which he claimed that variations in the beaks of these finches were caused by natural selection, and that Darwin's observations of these finch beaks had played a central role in the development of Darwin's big idea.[135]

---

[134] M.J. Behe, *Darwin Devolves: The New Science About DNA That Challenges Evolution* (New York, NY: HarperCollins, 2019).
[135] Wells, *Icons*, 162–163.

But all indications are that these finches did not actually play such a role. The finches are the subject of only one passing reference in the diary Darwin kept during his ocean voyage, and they are not mentioned at all in his book *Origin of Species*. Concerning the claim that the finches played a central role in the development of Darwin's big idea, historian of science Frank Sulloway has written, "Nothing could be further from the truth."[136]

Nevertheless, like so many other things concerning the theory of evolution, the myth that Galapagos finches were central to Darwin's thinking has been repeated so often—especially in biology textbooks—that these birds are now known universally as "Darwin's finches."

So what are Darwin's finches? They are often referred to as fourteen "species" of birds living on the Galapagos Islands or, in the case of one of the species, on the neighboring Cocos Island. The finches vary in body size and in the size and shape of their beaks. Some varieties have thin pointed beaks which are more useful in reaching nectar and small seeds in small spaces. Other varieties have beaks that are shorter and thicker, better suited for breaking hard nuts or other types of food.[137] The Galapagos finches are more similar to one another than to any species now on the mainland of South America, so it seems that a small group of common ancestors may have arrived on the islands long ago—perhaps about two million years ago—and changed slightly over time to produce the varieties now seen.[138]

If that is true, the Galapagos finches would, at most, be evidence for microevolution—but certainly not evidence for Darwin's big idea.

---

[136] Ibid., 161.
[137] Behe, *Darwin Devolves*, 144.
[138] Ibid., 146.

Figure 12. Galapagos finch species (plus one from the Cocos Island).

Such microevolution actually has been cataloged by a British husband and wife team from Princeton University, Peter and Rosemary Grant. The Grants probably know more about Darwin's finches than any other people on earth. They have conducted months long field studies of the finches on the Galapagos Islands annually since 1973. They have captured, tagged, and carefully measured finches each and every year—focusing mainly on one variety known as medium ground finches on one particular island.

The Grants have actually observed microevolution in progress. In 1977, the islands suffered a severe drought—resulting in the death of 85 percent of the medium ground finches. Varieties with thicker stronger beaks did better because they were able to crack the larger nuts of plants that had survived the drought. The Grants' measurements in subsequent years showed that finches that survived the drought had slightly bigger bodies and bigger beaks than the pre-drought population. "In other words, the Grants had meticulously documented evolution in the wild by variation, selection and inheritance—a feat that had eluded Charles Darwin himself—and very likely the same sort of process that accounted for all the differences among the Galapagos finch species."[139]

So if the Grants actually saw evolutionary changes in just a few decades, what amount of change had occurred in these geographically isolated finches during roughly two million years worth of that very same sort of evolutionary process?

> The results...produced by all that frenetic Darwinian evolution is a twofold variation in body length, shorter or longer beaks of greater or lesser depth, and not much else. Beginning with something very much like a finch, Darwinian processes labored long and mightily in the Galapagos and brought forth...a finch.[140]

Indeed, it is not even clear that all that evolution actually produced fourteen different species of finches. The Grants' careful observations for decades have produced many reports of interbreeding among finches thought to be of different species, which sometimes produces young more vigorous than their parents. Indeed, one "species" known as the large tree finch has apparently disappeared from one of the Galapagos Islands as the result of such crossbreeding. Such observations have produced an ongoing debate driven by some who now contend that what were thought to be fourteen species of finches may actually be not more than a handful of species.[141]

---

[139] Ibid., 145.
[140] Ibid., 146–147.
[141] Ibid., 147.

After two million years of natural selection, the finches are still finches. So Darwin's extrapolation from micro to macroevolution certainly did not occur in Darwin's finches. Why not?

To answer this question, we need to look beyond the bodies and beaks of birds and examine their genetics in great detail. Michael Behe starts by pointing out that the technology required for these important genetic studies has only recently been created.

> For the rigorous study of evolution, however, it's not nearly enough to just have knowledge of current biological processes—one has to be able to determine which particular mutations have occurred in individual organisms and what their effect has been…Experimental tools to sequence DNA developed only very slowly in the 1960s and 1970s and then explosively in the 1990s and later.

> One also has to examine huge numbers of organisms over many generations—or at the very least to examine the straightforward effects of mutations in modern populations whose history is well known. Only in the past twenty years have such detailed, rigorous evolutionary studies even begun to be conducted.[142]

Beginning in 1990, the Human Genome Project required about thirteen years to map the entire genetic code of a human. But by 2015, a team including the Grants was able in short order to map the entire genomes of *120 individual Galapagos finches*, including representatives from each of the so-called finch "species." Behe explains

---

[142] Ibid., 142.

what was shown by comparing the genes of blunt-beak birds to those of pointed-beak birds.

> [T[he gene that was statistically most strongly associated with differences between blunt-beak and pointed-beak finches [was] a gene called ALXi... ALXi codes for a protein consisting of 326 amino-acid units...Of those 326 positions a grand total of 2—count 'em, 2—differ between the pointiest-beaked finch and the bluntest... At position 112 in the blunt-beak gene there's a P instead of an L; at position 208 there's a V instead of an I. All other positions were identical. The letters that occur in the blunt gene [P and V] are the mutations...[143]

So what did that tiny genetic difference do to the protein produced by the ALXi gene? The research team reported that their computer analysis "classified both [changes] as damaging."[144] And how can a mutation that damages the normal function of a protein produce an advantage in the mutant bird?

> Well, if the normal activity of the ALXi protein during development helps make a beak sharper and more elongated, then hindering its activity could cause the beak to develop as less sharp and less elongated—in other words, shorter and blunter. If such beaks help a finch survive a drought, the mutant gene would be selected. Thus a beneficial mutation can be one that *damages* molecular machinery.[145]

---

[143] Ibid., 151.
[144] Ibid.
[145] Ibid., 151–152. (Emphasis original.)

Sequence analysis showed that the two mutations in the ALXi gene were not something that occurred in recent times but date back to soon after the birds arrived on the Galapagos Islands about two million years ago. Certainly other mutations had occurred during those two million years, but none of the other mutations had been preserved by natural selection. The blunt-beak mutation had helped survival during droughts, but it appears that no other mutation had helped survival in any significant way.

So what conclusion is to be drawn from the actual genetic record of Darwin's finches? Michael Behe put it this way:

> [M]illions of years of selection have left the finches very, very close to where they started… Surely we should expect one crummy new phylum, class or order to be conjured by Darwin's vaunted mechanism in the time the finches have been on the Galapagos. But no, nothing. A surprising but compelling conclusion is that *Darwin's mechanism has been wildly overrated—it is incapable of producing much biological change at all.*[146]

Microevolutionary Change by Genetic Damage Confirmed

Proponents of Darwin's big idea have long argued that geographic isolation should result in more evolutionary change, not less. The Galapagos finches are extremely isolated geographically because they are reproducing on small islands in the Pacific far from any continent. Yet the amount of evolutionary change observed in their actual genome is tiny.

Behe's analysis reinforces the conclusion about the very limited power of Darwinian mechanisms by describing a series of other geographically isolated creatures which show extremely limited evolutionary change. These include African cichlids (fish), Hawaiian

---

[146] Ibid., 155. (Emphasis added.)

fruit flies, Hawaiian beetles, Hawaiian honeycreepers (birds), and others.[147]

Isolation promoting change can be created by the hand of man as well as by geography. Dog breeding, for example, "isolates" certain breeding pairs for the purpose of creating change toward an intended result. Behe's analysis of genetic studies reports that well-known changes in dog breeds—including things like coat color, coat texture, overall size, tail length, and so on—are *all the result of the deletion of or damage to certain genes* in the dogs.[148]

Behe also reinforces his conclusion about the limited power of evolution by reporting the results of what he calls the most definitive evolution experiment ever conducted. This work was done by a microbiologist at Michigan State University.[149] Beginning more than twenty-five years ago, this researcher grew liquid cultures of common E. coli bacteria in laboratory flasks overnight. The next morning, he withdrew a 1 percent sample from the flasks and started brand-new cultures. He did this day after day after day for more than twenty-five years. Six to seven generations per day were produced for more than ten thousand days. All this amounted to more than sixty-five thousand generations of E. coli—the equivalent of more than a million years of human history.

The E. coli were allowed to do whatever came naturally. And to keep track of the genetic changes that occurred, these persistent workers sequenced the entire genomes of representative cultures after 500, 1000, 1500, 2000, 5000, 10,000, 15,000, 20,000, 30,000, 40,000, and 50,000 generations—for a total of 264 complete genomes.

> The bottom line is this. After fifty thousand generations of the most detailed, definitive evolution experiment ever conducted, after so much improvement of the growth rate that descendant cells leave revived ancestors in the dust,

---

[147] Ibid., 161–169.
[148] Ibid., 193–195.
[149] Ibid., 172–179.

after relentless mutation and selection, it's very likely that *all of the identified beneficial mutations worked by degrading or outright breaking the respective ancestor genes. And the havoc wreaked by random mutation had been frozen in place by natural selection.*[150]

And, of course, after fifty thousand generations of mutation and selection, the E. coli were still E. coli.

The Big Lesson: Major Differences between Creatures Require More Information, Not Less

The inherent limitation on the amount of change that can be produced by mutation and natural selection exists because mutations generally do damage—subtracting genetic information rather than adding it. This is why microevolution happens but is also why macroevolution can't. Genetic damage can sometimes be beneficial, but genetic damage can never produce new genetic information or new biological systems.

To fully appreciate this lesson, it is important to realize that very different kinds of creatures are actually made up of very different kinds of things. For example, fish, birds, and mammals are not simply differently shaped piles of the same kinds of cells. Instead, fish, birds, and mammals actually have different basic cell *types*. And research has shown that different cell types "require different, exceedingly complex molecular genetic regulatory networks."[151] Just as no one on earth has any idea how the complex mechanism for blood clotting might have evolved, no one even thoroughly understands how these different regulatory networks actually work, much less how they might have evolved.

The one thing we do know is that complex biological systems require complex genetic information to produce them. But genetic

---

[150] Ibid., 179. (Emphasis added.)
[151] Ibid., 159.

studies of Darwin's finches, and a series of other creatures, have now taught us that even advantageous mutations don't produce new information; they damage or destroy it.

So mutations plus natural selection cannot possibly explain how entirely new body plans might have arisen. To assume that the mechanisms of microevolution could by simple extrapolation explain the workings of macroevolution was both ignorant and wrong.

Behe puts the big lesson of *Darwin Devolves* this way:

> [R]andom mutation and natural selection both promote evolution on a small scale and hinder it on a large one. Mutation supplies the variation upon which natural selection acts, but the greatest amount of that variation comes from damaging or outright breaking previously working genes. In the case of an already functioning complex system, natural selection shapes it more and more tightly to its current role, making it less and less adaptable to other complex roles.[152]

> From the beginning the Darwinian mechanism has been self-limiting, capable to an extent of eliminating or modifying preexisting molecular systems and in the process giving rise to new varieties of creatures below the biological classification level of family, *but incapable of building functionally complex molecular structures.*[153]

In his book *The Wedge of Truth*, Phillip Johnson wrote at length about the fact that mutations plus natural selection cannot possibly produce the new information required for macroevolution to be a reality.

---

[152] Ibid., 246.
[153] Ibid., 250–251. (Emphasis added.)

He begins by describing a 1998 interview of the person who is perhaps the world's most famous Darwinist, Richard Dawkins, in which Dawkins was asked whether he could give a single example of a mutation or other evolutionary process that was information-enhancing. Johnson reports that "Dawkins hesitated for at least eleven seconds, an agonizingly long time in the context of a video interview, before he finally gave a completely irrelevant reply about the transition between fish and amphibians."[154] Dawkins quickly published a follow-up article to cover his failure to answer the question about information. The paper speculated "about how a hemoglobin molecule might have evolved from a predecessor through random mutations in inactive genes…but there [was] no explanation of how mutations in genes which are inactive (and hence not subject to natural selection) can be causing massive increases in genetic information. Above all, there [was] no description of any mutations which are actually known to have the kind of information-creating power which would be required for creative evolution."[155]

In other words, even the world's leading advocate for Darwin's big idea can't answer the most basic question about the source of the more and more and more genetic information required for increasingly complex forms of life. And we should not blame him for being at a loss for words because even the most recent genetic research shows that there is no good answer to the question.

The fundamental problem has been that evolutionists have acted as if information and matter are the same thing, but they are not. This error was recognized in 1995 by an award-winning evolutionary biologist named George C. Williams.

> Evolutionary biologists have failed to realize that they work with two more or less incommensurable domains: that of information and that of matter… The gene is a package of information, not an object. The pattern of base pairs

---

[154] Johnson, *The Wedge of Truth*, 39–40.
[155] Ibid., at 41–42.

in a DNA molecule specifies the gene. But the DNA molecule is the medium, it's not the message. Maintaining this distinction between the medium and the message is absolutely indispensable to clarity of thought about evolution.[156]

So genetic copying errors can effect matter, but they can never produce the information needed for the development of new forms of living matter. For this further reason, Darwin's big idea is contrary to the evidence.

---

[156] Johnson, *Defeating Darwinism*, 70, quoting George Williams in *The Third Culture: Beyond the Scientific Revolution*, ed. John Brockman (New York: Simon & Schuster, 1995), 42–43.

CHAPTER 6

EVEN 4.5 BILLION YEARS IS NOT NEARLY ENOUGH TIME

Evolutionists and creationists have fought for a very long time about the age of the earth, but that dispute is nothing more than a great distraction of no real consequence to the problems with Darwin's big idea. Actually, only some of those who believe in divine creation believe the earth is young. Other creationists, often called "old earth creationists," agree with evolutionists that life on earth is a few billion years old.[157] Guesstimates range up to perhaps 4.5 billion years for the age of the oldest life on earth.

The important point is that *even if life on earth is billions of years old, there still has not been nearly enough time for macroevolution to have produced the millions of forms of life we now see.*

Darwin's big idea was that something like one living cell (or maybe a handful of different living cells), by a random undirected process of mutation and natural selection, slowly developed into all the living forms of life we see on earth. He believed this process to be exceedingly slow because he was sure the changes could occur only by an unimaginably long series of tiny steps. He used words like "interminable" and "infinitude" and "infinite" to describe the number of changes that must have occurred for that single cell to become, for example, a human.[158]

---

[157] See Reasons to Believe at https://reasons.org/.
[158] Darwin, *Origin*, 438 and 453.

A Few Billion Is a Lot of Years, but It's Still Not Nearly Enough

What is the mathematical probability that macroevolution actually occurred? Evolutionists like to assume, and sometimes even say, that billions of years would have been plenty of time for such an "infinitude" of changes to occur. In effect, the argument is sometimes made that given billions of years, "anything can happen."

Comments like one by Darwin apologist Richard Dawkins are common: "Given, say, a hundred million Xs, we should be able to construct a plausible series of tiny gradations linking a human eye to just about anything!"[159] For another similar example, in 1954, Harvard biochemist George Wald argued in the context of a couple of billion years, "given so much time the 'impossible' becomes possible, the possible probable, and the probable virtually certain. One has only to wait: time itself performs the miracles."[160]

But does it? Actual mathematics—in the form of something called probability science—shows convincingly that it does not.

I'm no mathematician, and maybe you aren't either. But I can read the writings of people who are very skilled with mathematics and can provide us with understandable narrative descriptions of mathematical realities. These folks know that the probability numbers required for macroevolution to have occurred are way too big to have occurred on earth—even if life on earth is a few billion years old.

Nevertheless, students are told repeatedly that even the most improbable events can occur in a few billion years. For example, in a college-level biology text cowritten by late evolutionary biologist and Harvard professor Stephen Jay Gould, it was stated on three separate occasions that given enough time, it is possible to flip a coin

---

[159] Dawkins, *The Blind Watchmaker*, 78.
[160] Johnson, *The Wedge of Truth*, 50, quoting George Wald, "The Origin of Life," *Scientific American*, August 1954, 44–53.

one hundred heads in a row.[161] But this famous author was badly mistaken:

> Their statement is misleading on the time-scale of the universe itself. If you flipped an honest coin once a second continuously around the clock, then you would require the number 200 thousand billion times the maximum estimated age of the universe [which is much older than life itself] to flip a trial of 100 heads. This is no exaggeration. On average it would require that much time. This narrative description explains the extremely low probability in terms we can humanly grasp.[162]

Another world-famous man has made a very similar mistake about the actual amount of time needed for the occurrence of a very improbable event. In his book *A Brief History of Time*, physicist Stephen Hawking repeated with approval the classic saying that if one assembled "a horde of typing monkeys," then, according to Hawking, "very occasionally by pure chance they will type out one of Shakespeare's sonnets."[163]

Again, mathematics proves this idea to be false. It is still false even if the problem is vastly reduced to the probability of only the first one hundred characters of a sonnet being typed by an impossibly large number of monkeys typing much faster than the best human typist can type, for a time much longer than the age of the universe.

> If each proton in the observable universe were a typing monkey (roughly 10^{80} in all), and they

---

[161] S.J. Gould, S.E. Luria, and S. Singer, *A View of Life* (Menlo Park, CA: The Benjamin/Cummings Publishing Company, Inc., 1981), 592, 689, and 693.

[162] Walter James Remine, *The Biotic Message: Evolution versus Message Theory* (St. Paul, MN: St. Paul Science, 1993), 75.

[163] S.W. Hawkins, *A Brief History of Time: From the Big Bang to Black Holes* (New York: Bantam Books, 1988), 123.

> typed 500 characters per minute (faster than the
> fastest secretary), around the clock for 20 billion
> years, then all the monkeys together could make
> 5×10^{96} attempts at the 100 characters. It would
> require an additional 3×10^{46} such universes to
> have an even chance at success. We scientifically
> conclude that the monkey scenario cannot suc-
> ceed. For the scientist it would be perverse to
> insist otherwise.[164]

The human genome is way longer than a full Shakespearian sonnet. Rather than consisting of a few hundred letters, each cell of the human body contains about *three billion* DNA base pairs, containing more than *six billion* letters of code. This is the complexity of the biological "blueprint" supporting human life.

How long would it take for the random strikes of evolutionary monkeys to type that blueprint with the precision necessary for life? Here's how one MIT physicist has described the impossibility—focusing not on the random assembly of an entire living creature but merely the assembly of only one relatively small protein:

> Proteins vary in length from strings of a few hun-
> dred to a few thousand amino acids. Consider a
> relatively short protein, such as 200 amino acids
> long. Into each of the 200 spaces any one of the
> 20 amino acids can fall. That means the total
> number of possible combinations is 20 times
> 20 times 20 repeated 200 times. The result is a
> one with 260 zeros after it, or 10 to the power of
> 260 or a billion billion billion repeated 29 times.
> From this vast biological grab bag, we are told
> that nature by random chance mutations has
> been able to form the fewer than a million pro-
> teins that don't kill the organism. This did not,

---

[164] Remine, *The Biotic Message*, 80.

and could not, and will not happen by chance. And every biologist enamored with neo-Darwinian evolution knows it.[165]

In Theory, How Many Generations of a Creature Would Be Needed to Produce Significant Macroevolutionary Change?

Darwin's big idea was that a mutation would produce some survival advantage in a parent, the parent would then pass along that advantage to its child, and that additional beneficial changes in future generations would eventually result in some entirely new biological mechanism. Eventually, after sufficient repetition of this process, an entirely different type of creature would result.

So how many generations of a creature would be required to produce such change?

No one really knows the answer to this question, but educated guesses can be made based on computer modeling. As you might expect, the results of such models depend in very large part on the assumptions that are built into the model. These include assumptions about the proportion of mutations that are neutral versus damaging, the number of different genetic locations at which the beneficial change could occur, the number of beneficial changes required, the size of the population of organisms, and so on.

In 2004, Michael Behe and a colleague published an article reporting the results of just such a computer modeling project. They focused not on the development of an entire biological "mousetrap" but merely on the development of one small piece of such a thing. Think of, say, the mousetrap's spring. They called these "mini-irreducibly complex features" or "mICs." The goal of their work was not to estimate the number of generations required to produce any particular mIC but rather to estimate how many *more* generations

---

[165] Gerald Schroeder, When Pigs Fly, and Monkeys Type (2007), Catholic Education Resource Center, reprinted from tothesource.com and available at https://www.catholiceducation.org/en/science/faith-and-science/when-pigs-fly-and-monkeys-type.html.

would be required if a particular mIC required two, three, four, or more mutations rather than just one mutation.[166]

Immediately upon publication of Behe's article, a mathematical geneticist named Michael Lynch published his own article in the very same journal in an effort to rebut Behe's conclusions. Lynch made entirely different assumptions—all of which favored the idea of an mIC resulting from a smaller number of generations. Behe describes this back and forth as "pretty close to the ideal of science" in testing his work because "if the strongest criticisms of knowledgeable and dedicated opponents don't topple it...its credibility is strengthened."[167]

While the competing papers reached different conclusions about the number of generations needed to develop a particular mIC, they agreed on a very important general principle. Behe describes it this way:

> [I]t's expected to take about ten thousand generations to mutate just one particular amino acid in a particular protein. To produce a feature... that needs two such mutations, in Lynch's model a hundred million generations are needed. Our model indicates about a billion. A hundred million is of course much less than a billion, but both numbers are much, much greater than [the ten thousand generations] needed for a single mutation.[168]

> If just two simple molecular changes are needed for a feature to evolve, there's a quantum leap in difficulty for Darwin's mechanism. The more

---

[166] Behe, *Darwin Devolves*, 238–240.
[167] Ibid., 240.
[168] Ibid., 241.

required changes, the exponentially worse it becomes.[169]

Remember, both researchers were talking only about the number of generations required to develop the biological equivalent of just the spring of a mousetrap—and they were assuming that it's only a two-step process, which is far too short. How many steps and how many generations would be required to develop the whole mousetrap? Or a system as complex as blood clotting? Or sexual reproduction?

And because the problem grows "exponentially" as the number of steps increases, we are really talking about not just a hundred million or so generations but billions or trillions of generations for truly complex developments. Trillions of generations might be conceivable if you are a bacteria or a fruit fly but certainly not if you are a cow or a person. For us, a few billion years is nowhere near enough time to produce many billions of generations.

Michael Denton recently summarized the problem this way:

> What has been discovered at the heart of life is what I have previously termed the "third infinity." Whereas the cosmos is the infinity of the very large, and the atom is the infinity of the very small, the organism is the infinity of the very complex. That such an infinity might have come about *in finite time* as a result of any sort of undirected random process seems impossible.[170]

Evolutionists have long known about but simply dismissed the problem of mathematics showing there has not been enough time on earth for macroevolution to have occurred as Darwin supposed. A well-known mathematician made such a claim at a conference on evolution in 1967, but his numbers were disregarded because the

---

[169] Ibid., 242.
[170] Denton, *Evolution: Still a Theory in Crisis*, 226. (Emphasis original.)

evolutionists were unwilling to consider the possibility that Darwin's big idea was wrong.

> [The mathematician] argued that it was highly improbable that the eye could have evolved by the accumulation of small mutations, because the number of mutations would have been so large and the time available was not nearly long enough for them to appear. [Evolutionists] argued that [the mathematician] was doing his science backwards; the fact was that the eye had evolved and therefore the mathematical difficulties must only be apparent.
>
> [I]t was as if [the mathematician] had presented equations proving that gravity is too weak a force to prevent us all from floating off into space. Darwinism to them was not a theory open to refutation but a fact to be accounted for...[171]

As usual, the Darwinian story was far more important than the numbers. But common sense tells us that numbers are often far more reliable than storytellers.

---

[171] Johnson, *Darwin on Trial*, 38–39.

CHAPTER 7

WHY IS BIOLOGY FULL OF TRAITS THAT DO NOT PROMOTE SURVIVAL?

Only Means Only, Doesn't It?

Way back in chapter 1, I noted that Darwin himself had said that if his big idea were true, there would be certain other things that must be false. One of those was that no organ or organ system could exist that could not have developed by an infinite series of gradual, tiny steps.[172] But we have not yet discussed another fact of nature that Darwin said could not exist if his big idea were true.

As you now well know, Darwin's big idea was that natural selection preserves only those traits in a population of organisms that help those organisms survive and reproduce. Completely consistent with this basic idea, Darwin also insisted that "natural selection will *never produce in a being anything injurious to itself*, for natural selection acts *solely by and for the good of each*."[173]

Notice that Darwin went well beyond the obvious point that natural selection would not preserve anything that actually made it *less likely* that an organism would survive and reproduce. He said that natural selection "acts *solely* for the good of each," meaning that natural selection would not preserve anything that did not actually

---

[172] Darwin, *Origin*, 219. "If it could be demonstrated that any complex organ existed, which could not possibly have been formed by numerous, successive, slight modifications, my theory would absolutely break down."

[173] Ibid., 229. (Emphasis added.)

benefit an organism. This means that every single trait of every single organism should actually promote survival and/or reproduction.

So is that the case? Far from it.

Nature Is Full of Traits That Have No Adaptive Purpose

We have been taught to believe that "survival of the fittest" is a very powerful idea. Since Darwin's big idea says that natural selection only "selects" things that actually help the creature survive and reproduce, we often assume that everything about a living organism plays some part in helping it survive and reproduce. While it is easy to identify certain traits of living things that serve these purposes, the fact is that there are many, many traits that don't.

In his 2016 book *Evolution: Still a Theory in Crisis*, Michael Denton put it this way:

> The fact that many [traits] exhibit curious geometric and numeric features reinforces the impression that they are indeed abstract non-adaptive patterns, quite beyond the explanatory reach of any adaptationist or selectionist narrative.
>
> A major problem in defending the Darwinian claim…is the existence of a vast universe of non-adaptive forms and patterns in nature which no biologist—*not even the most convinced functionalist or Darwinist*—has ever viewed as adaptive.[174]

Some of these could not be more simple. Consider the shape of the leaves on your favorite tree. Why does an oak leaf have the shape it does? Or a birch? Or a maple? Survival of the fittest? Really? When they are growing right next to each other in the very same woods?

---

[174] Denton, *Evolution: Still a Theory in Crisis*, 70 and 76. (Emphasis original.)

Figure 13. Oak, birch, and maple leaves.

Or why do many plants have leaves emerging from a stem directly opposite from one another while many others have leaves emerging alternately? How does each pattern promote survival? Why don't random mutations result in the random placement of leaves?

Opposite   **Alternate**

Figure 14. Twigs showing typical opposite placement of leaves as contrasted with typical alternate placement.

Some other examples are not so simple.

One of these is the pattern and configuration of the bones in the skull of a human infant at birth. As shown in figure 15, rather than being a solid globe, five skull bones are joined together by fibrous structures covering openings in the bones known as fontanelles. This allows the skull to flex and thereby ease the movement of the skull through the cervix during birth.

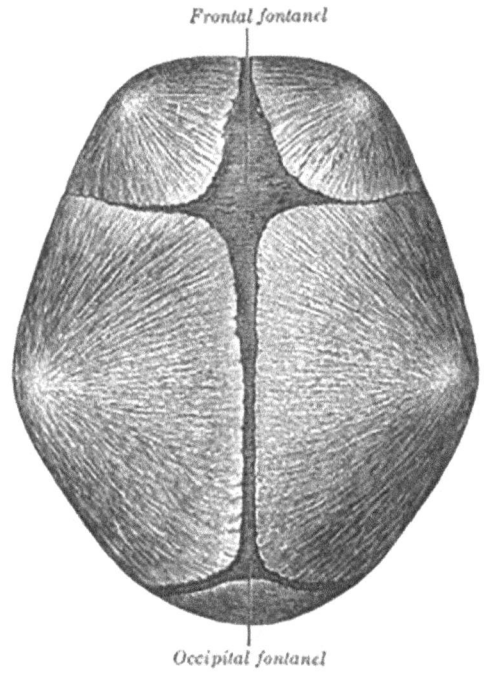

Figure 15. The human skull at birth, showing the anterior (frontal) and posterior (occipital) fontanelles.

Yes, that definitely sounds adaptive. But here's the problem: a very similar pattern of skull bones exists in birds, reptiles, and fish. These creatures hatch from eggs rather than navigate a birth canal. So why has natural selection "selected" and preserved this pattern for the skull of a chick pecking its way out of an egg? Or a tiny snake using its egg tooth to slash its way out?[175]

Denton describes an almost never-ending series of nonadaptive traits—each of which challenges the random mindlessness of Darwin's big idea.

> Why would virtually every mammal on earth—whether mouse or giraffe—have exactly seven vertebrae in its neck?[176]

> Why would every insect on earth (at least one million species) have exactly three body segments—head, thorax, and abdomen—and never more or less?[177]

> Why would every insect limb have the same five segments: coxa, trochanter, femur, tibia and tarsus?[178]

> Why would every one of more than three thousand species of centipedes on earth have an odd number of body segments, never an even number?[179]

> Why does the forelimb of every vertebrate tetrapod [four-limbed creature; see figure 17 below] follow the exact same 1-2-5 (pentadactyl) bone pattern regardless of whether it is the wing of a bat, the flipper of a whale or the arm of a human?[180]

---

[175] Ibid., 66–67.
[176] Ibid., 70.
[177] Ibid., 47.
[178] Ibid.
[179] Ibid., 50.
[180] Ibid., 45–46.

> Why do all of the flowers of seed-bearing plants—
> an extremely diverse group of at least 300,000
> species—have exactly the same pattern of con-
> centric whorls: carpels within stamens within
> petals within sepals?[181]

And the list of examples goes on and on and on. Denton summarizes the situation with the simple conclusion that "a great deal of organic order appears to be non-adaptive."[182]

Darwin was very much aware of the mystery of why nonadaptive traits would be preserved by natural selection, but he just dismissed it out of hand. He insisted that it was reasonable to "suppose" that some ancient ancestor had a particular pattern for some now unknown reason and that pattern was somehow preserved even though it continued to serve no identifiable purpose. But nowhere in *Origin of Species* did he provide any solid justification for this huge supposition. "In other words, [Darwin] effectively ignores the profoundly subversive fact that a great deal of order in living organisms has never been shown to be adaptive either in extant [existing] or in ancient forms."[183]

Based on the widespread reality of nonadaptive organic forms, Michael Denton draws a profound conclusion.

> No matter how many times Darwinists reiterate
> the fairy story that [widespread non-adaptive
> patterns] were "once upon a time" adaptive in the
> ancestor of the clade [group] they define, it is a
> claim without the slightest empirical or rational
> basis...
>
> If...the apparently non-adaptive [patterns] which
> underlie the adaptive diversity of life are indeed

---

[181] Ibid., 48–49.
[182] Ibid., 82.
[183] Ibid., 75.

non-adaptive, then the whole Darwinian edifice stands on sand, on an unproven assertion which can never be finally proved, and which seems exceedingly unlikely to be true.[184]

Nature Surprises Us with Altruism for No Apparent Purpose

In 1850, British poet Alfred Tennyson published a famous work after the sudden death of a friend which described nature as being blood "red in tooth and claw." When *Origin of Species* was published less than a decade later, some recalled that phrase as an apt description of Darwin's notion of survival of the fittest. Modern Darwinist Richard Dawkins used that phrase in his modern classic *The Selfish Gene* to describe the ruthless competition between living things.

Aggression, of course, is not an organ but a behavior. Nevertheless, Darwin thought his big idea applied to behaviors as well. Chapter 7 of *Origin of Species* was titled "Instinct," which imagined that "mental powers"—sometimes called habits or instincts— were subject to the same developmental dynamics of variation and natural selection as were physical forms.

In his summary of chapter 7, Darwin concluded that the behaviors of life were controlled by self-interest just like everything else. He wrote, "[N]o instinct has been produced for the exclusive good of other animals, but that each animal takes advantage of the instincts of others…"[185] This thought was very similar to the prediction noted at the beginning of this chapter—that natural selection will only preserve things, including behaviors, that are of benefit to the owner of that behavior.

But is this all we see in nature? Remember, Darwin said that natural selection would act "solely" for the good of the organism itself.

Consider the scene in figure 16 showing a crocodile and a bird known as a plover.

---

[184] Ibid., 75–76.
[185] Darwin, *Origin*, 263.

Figure 16. An Egyptian crocodile opens wide for a tooth cleaning by a plover.

The plover is a bird that normally wades in water and feeds on insects. But apparently the plover is also known to feed on decaying meat that it plucks from the teeth of the Egyptian crocodile. At least at first blush, it looks like the opposite of blood "red in tooth and claw" survival of the fittest, doesn't it?

In fairness, we should recall that Darwin's claim was that there would never be a behavior that was "for the *exclusive* good of the other animal." In theory, perhaps the crocodile might benefit somehow from the tooth cleaning. But this also seems highly unlikely because crocodiles are polyphyodonts—meaning that older, worn teeth are constantly being replaced. Perhaps even more importantly—since every evolutionist insists that natural selection is completely mindless and lacks any ability to plan ahead—it is hard to see how the prospect of tooth decay tomorrow would stop a crocodile from having a plover for lunch today.

So is this scene a Darwinian impossibility—a case of the Egyptian crocodile developing a behavior that is exclusively for the benefit of the plover? Maybe.

But if you are still wondering about this one, consider the many reports of creatures *adopting the young of different species*.

Inter-Species Adoptions

The key here is that the beneficiaries of this kindness are member of *different species*. This reality totally negates the easy explanations of Darwinists whenever they see unexpected kindness in nature. These favorite explanations rely on the concepts of kinship or reciprocal kindness. Kinship explains why a parent or sibling might give its life for another—because that other actually has some of the very same genes as the one making the sacrifice. Reciprocal kindness says that I might scratch your back expecting that sometime later, you might scratch mine.

You might quickly see that the explanation of reciprocal kindness suffers from that common problem of requiring the one giving up time and energy for another to be "looking down the road" to a possible future benefit for himself—something we are told repeatedly that mindless natural selection cannot do.

In any event, neither of these easy explanations can apply where the organism making the sacrifice and the one receiving the benefit are members of different species. Obviously, such a pair do not share the same genes. And even if the one making the sacrifice were inclined to consider the possibility of receiving some future benefit in return, the members of different species have no reasonable expectation that, in the future, they will be living in the same community of creatures or paying any attention to one another at all.

So how common are inter-species adoptions? More common than you might think. Field biologists observe it quite frequently and have reported their observations in many published articles. It has been observed in many different bird species—falcons,[186] robins,[187] crows,[188]

---

[186] Alexandre Anctil and Alastair Franke, "Intraspecific Adoption and Double Nest Switching in Peregrine Falcons," *Arctic Institute of North America* 66, no. 2 (2013): 222–225, https://www.jstor.org/stable/23594687.

[187] Asa Berggren, "Intraspecific Adoption and Foster Feeding of Fledglings in the North Island Robin," *New Zealand Journal of Ecology* 30, no. 2 (2006): 209–217, https://www.jstor.org/stable/24056341.

[188] Kevin J. McGowan, "Nesting Crows Adopt a Fledgling Blue Jay," *Journal of Field Ornithology* 62, no. 2 (1990): 171–173, https://sora.unm.edu/sites/default/files/journals/jfo/v061n02/p0171-p0173.pdf.

terns,[189] warblers,[190] petrels,[191] sparrows,[192] and eagles,[193]—as well as in chimpanzees,[194] cheetahs,[195] cichlids (fish),[196] and even in ants.[197]

[189] Stephen A. Oswald, Christy N. Wails, Brittany E. Morey, and Jennifer M. Arnold, "Caspian Terns Fledge a Ring-Billed Gull Chick: Successful Waterbird Adoption Across Taxonomic Families," *Waterbirds* 36, no. 3 (2013): 385–389, http://www.bioone.org/doi/abs/10.1675/063.036.0318.

[190] Cameron J. Fiss, Darin J. McNeil, Renae E. Poole, Karli M. Rogers, and Jeffrey L. Larkin, "Prolonged Interspecific Care of Two Sibling Golden-Winged Warblers by a Black-and-White Warbler," *The Wilson Journal of Ornithology* Vol. 128, no. 4 (2016): 921–926, http://wjoonline.org/doi/abs/10.1676/15-180.1?code=wors-site.

[191] Terrence W. O'Dwyer, Dean Portelli, and Nicholas Carlile, "Interspecific Fostering of a Wedge-Tailed Shearwater by White-Necked Petrels on Phillip Island, Norfolk Island Group," *Marine Ornithology* 46, (2018): 43–45, https://www.researchgate.net/profile/Nicholas_Carlile/publication/325333562_Interspecific_fostering_of_a_Wedge-tailed_Shearwater_Ardenna_pacifica_by_White-necked_Petrels_Pterodroma_cervicalis_on_Phillip_Island_Norfolk_Island_Group.

[192] G.A. Lozano and R.E. Lemon, "Adoption of Yellow Warbler Nestlings by Song Sparrows," *The Wilson Bulletin* 110, no. 1 (1998): 131–133, https://pdfs.semanticscholar.org/23b6/fb986610a773129d633835d556c5a8b3a071.pdf.

[193] Robert G. Anthony and John T. Faris, "Observations of a Live Glauous-Winged Gull Chick in an Active Bald Eagle Nest," *The Wilson Bulletin* 115, no. 4 (2003): 481–483, https://www.jstor.org/stable/4164612.

[194] Christophe Boesch, Camille Bole, Nadin Eckhardt, and Hedwin Boesch, "Altruism in Forest Chimpanzees: The Case of Adoption," PLOS (2010), https://journals.plos.org/plosone/article?id=10.1371/journal.pone.0008901.

[195] Sarah M. Durant, Sultana Bashir, Thomas Maddox, and M. Karen Laurenson, "Relating Long-Term Studies of Conservation Practice: the Case of the Serengeti Cheetah Project," *Conservation Biology* 21. no. 3 (2007): 602–611, https://pdfs.semanticscholar.org/a029/480e4629a4ecfbc9bee59bd09d8a9d255abf.pdf.

[196] Brian D. Wisenden and Miles H.A. Keenleyside, "Intraspecific Brood Adoption in Convict Cichlids: A Mutual Benefit," *Behavioral Ecology and Sociobiology* 31, (1992): 263–269, http://web.mnstate.edu/wisenden/reprint%20pdfs/1992%20Brood%20adoption%20BES.pdf; and Franziska C. Schaedelin, Wouter F.D. van Dongen, and Rishard H. Wagner, "Non-Random Brood Mixing Suggests Adoption in a Colonial Cichlid," *Behavioral Ecology* 24, no. 2 (2013): 540–546,.

[197] Kathleen P. Rudolph and Jay P. McEntee, "Spoils of war and peace: enemy adoption and queen-right colony fusion follows costly intraspecific conflict in

My own personal experience with animal adoption has taught me that it is not an easy thing to accomplish. My experience dates back to the days of my youth on a sheep farm trying to care for a lamb that had been rejected—not allowed to nurse—by its own mother. We would try to convince another ewe with only one lamb to allow the orphaned lamb to nurse, but our efforts would usually fail. Every nursing mother seemed to know immediately by smell whether a lamb was her own and if not, she would not let it nurse.

So if adoption is so challenging within the same domesticated species, I was especially surprised to learn that adoptions actually occur in the wild—and across species, no less!

The undeniable existence of a multitude of nonadaptive traits in nature raises a very big question. Since Darwin's big idea cannot explain this reality, what can? Even without knowing that answer, we must conclude that something other than evolution—something very powerful—is going on in nature.

Before Darwin's big idea silenced most critics, some very famous biologists had some very well-developed thoughts about the nature of this other powerful force. One of these was the founder of London's Museum of Natural History and a contemporary of Darwin, Richard Owen.

> Owen believed that there was a substantial degree of order inherent in living systems, manifest in what he termed "primal patterns," the grand taxa-defining [patterns] or ground plans that underlie the adaptive diversity of life. Owen argued that many of these ground plans (like the pentadactyl ground plan of the tetrapod limb) do not appear to be adaptive. That is, they do not appear to have or to ever have had any role in fashioning actual organisms to meet specific adaptive ends. Such "primal patterns"...there-

acacia ants," *Behavioral Ecology* 27, no. 3 (2016): 793–802, https://academic.oup.com/beheco/article/27/3/793/2365688.

fore could not be the result of adaptive evolution as Darwin claimed. Owen believed...that these deep homologous patterns were immanent aspects of the world order which arose in some way from the intrinsic physical properties of living things.[198]

Owen explained his nonadaptive theory of life in a landmark treatise called *On the Nature of Limbs*, published in 1849. The point here is not to understand his theory in detail. The point here is simply to realize that before Darwinism shut down debate, many very thoughtful biologists had many very interesting ideas about the actual nature of life.

One thing we can know for sure is that Darwinian evolution falls far short of explaining everything we see in nature.

---

[198] Denton, *Evolution: Still a Theory in Crisis*, 61–62.

CHAPTER 8

WHY ARE BIOLOGY TEXTBOOKS FULL OF MISINFORMATION ABOUT EVOLUTION?

Since many biologists have been talking about "primordial soup" and "missing links" now for more than a century and a half, it must seem obvious to textbook authors that they have to say something about Darwin's big idea that every plant and animal on earth was somehow produced by mutations and natural selection. This is especially true when this idea has been declared official scientific truth by umpteen professors, including that elite scientific club known as the National Academy of Sciences. Indeed, by the 1970s, the unmitigated praise for Darwin's big idea had become so extreme that one Darwinist felt confident enough to title an article written for biology teachers "Nothing in Biology Makes Sense Except in the Light of Evolution."[199] If that's the case, then the first goal of every biology textbook must be to make sure that each student thinks the theory of macroevolution is true.

So what does our textbook author do if he or she has no real, solid evidence that Darwin's big idea is true? Judging from one group of ten biology texts, the answer is to repeat without analysis a series of "icons of evolution" regardless of whether they were disproven decades ago or even known to have been frauds in the first place. I know this must sound like a vast overstatement of what's been happening, but it's not.

---

[199] Theodosius Dobzhansky, "Nothing In Biology Makes Sense Except in the Light of Evolution," *The American Biology Teacher* 35, (1973), 125–129.

In the year 2000, molecular biologist Jonathan Wells published a book called *Icons of Evolution: Science or Myth?*[200] The book considers ten of the most common ideas cited in support of Darwin's big idea—the ten being what Wells calls the "icons of evolution"—and judges whether or not they are supported by the evidence. He comes to the firm conclusion that they are not—not a single one. Remarkably, he shows that more than one of these "icons" was actually fabricated by the author of the icon. In an appendix to the book, Wells "grades" ten leading biology textbooks on how well they describe the truth about seven of the ten icons. Of these seventy grades—ten books each discussing seven icons—sixty-eight of the seventy grades are either a "D" or an "F." Not a pretty picture.

Less Science Than Myth

I began this little book by betting that you have been lied to about the theory of evolution. I felt confident in making that bet because I know that ten of the most popular ideas used to promote Darwin's big idea are false. Because the use of these icons is so widespread, I want to take the time to briefly review what Wells has to say about them so you will better understand the lies you've been told.

As you read these summaries, please keep this in mind: Jonathan Wells has been criticized loudly by the "scientific community." These mainstream biologists have treated every other critic of Darwinian evolution with exactly the same disrespect, including many of the other authors I have cited. These Darwinists say things like Jonathan Wells' thoughts have been "widely rejected by the scientific community." But I have read no criticism of Wells that explains *why* his thoughts have been rejected, much less the good reasons for such rejection. As usual, proponents of evolution rely on opinions, not evidence.

Wells explains in great detail the evidence undercutting the credibility of each of the icons of evolution. Since I will present only a brief summary of each explanation, I will be relaying only a very small part of

---

[200] Jonathan Wells, *Icons of Evolution: Science or Myth?* (Washington, DC: Regnery Publishing, 2000).

this evidence and relying largely on simple logic. I will try to use points that seem to be true beyond all reasonable doubt. Let common sense be your guide rather than the conclusory opinions of the "experts."

1. *The Miller-Urey Experiment*

Darwin did not purport to know how life arose in the first place but merely how that original form of life may have developed into the millions of different forms of life we see on earth in modern times. Indeed, as noted above, Darwin guessed that all life on earth may have "descended from some one primordial form, into which life was first breathed."[201]

Evolutionists, nevertheless, have realized for a long time that having no explanation whatsoever for how life first arose leaves a gaping hole in their naturalistic worldview. So in the 1920s, a couple of scientists speculated that the ancient earth may have contained the "warm little pond" of chemicals imagined by Darwin which, when struck by lightning, may have produced some original building blocks of life.

In 1952, a pair of chemists named Stanley Miller and Harold Urey conducted an experiment that purported to simulate the conditions of ancient earth and produced certain amino acids, considered to be basic building blocks of life. Published in 1953, their work quickly found its way into most high school and college biology textbooks and became an important part of the argument that scientists used to demonstrate how life may have arisen naturally on earth. By this means, a major hole in Darwin's natural explanation of the living world was thought to have been filled.

But was it? It is important to remember that the Miller-Urey experiment was conducted by scholars in a university laboratory using a relatively complicated set of equipment—none of which existed on earth eons ago. So the only basis that might exist for thinking this experiment produced useful information about the origin of life is the extent to which the experimenters actually succeeded in recreating the actual conditions that existed on earth so long ago.

---

[201] Darwin, *Origin*, 455.

The Miller-Urey apparatus involved a closed circle of glass tubing and glass globes from which normal air was removed and replaced with other gases—a combination of methane, ammonia, and hydrogen, but no oxygen (which would have been explosive). Water was heated in this closed system and the humid gases were circulated past a high voltage electric spark to simulate possible lightning. The glass tubing also included a glass trap—much like the trap under your bathroom sink—to collect anything that may have settled out from the liquid. After a number of days, the water became red and cloudy. When the material in the trap was examined, it was found to include certain organic compounds known as amino acids, which, if constructed into long chains as directed by DNA, could form the proteins of living organisms.

The first thing to be noted about the Miller-Urey experiment is that while it produced certain amino acids, *it did not produce life*. So it is extremely misleading to say that the experiment is evidence supporting the origin of life. At most, it is evidence supporting the origin of amino acids. As Jonathan Wells has pointed out, "A biochemist can mix all the chemical building blocks of life in a test tube and still not produce a living cell."[202]

Citing a few other honest scientists, Wells has argued that the experiment was further flawed because it relied on atmospheric conditions in its glass tubes that were very much *unlike* the actual conditions on earth billions of years ago.

- The experiment excluded oxygen from its glass tubes because, in the early 1950s, it was assumed that the ancient atmosphere contained no oxygen. But it is well-known that water was abundant on the ancient earth's surface and that its components, oxygen and hydrogen, are easily split by ultraviolet light from the sun. So from the 1960s through the 1980s, a debate raged as to how much oxygen may have existed in the ancient atmosphere.[203]

---

[202] Wells, *Icons*, 24.
[203] Ibid., 14–18.

- By 1982, a conference of researchers on the origin of life decided that there must not have been oxygen in the earth's ancient atmosphere because any organic compounds that may have been created could not have survived in such an atmosphere.[204] Yet that same year, a pair of British geologists published evidence showing "from the time of the earliest dated rocks at 3.7 billion years ago, Earth had an oxygenic atmosphere." They said it was mere "dogma" to claim that the earth's early atmosphere lacked oxygen.[205]

- By 1996, a paleobiologist at the Smithsonian Institution had reviewed the evidence and concluded that "the Earth very likely had an atmosphere that contained free oxygen."[206]

- Independent of the debate about the presence of oxygen, however, the atmospheric assumptions of the Miller-Urey experiment still fail. In 1966, another paleobiologist asked and answered, "What is the evidence for a primitive methane-ammonia atmosphere on Earth? The answer is that there is *no* evidence for it, but much against it."[207]

- In 1983, Stanley Miller had himself reported that repeating his experiment in the absence of methane gas produced only a small amount of one amino acid, glycine.[208]

---

[204] Ibid., 18. Yes, perfectly circular reasoning such as this is extremely common among evolutionists.

[205] Ibid., 18, quoting Harry Clemmey and Nick Badham, "Oxygen in the Precambrian atmosphere: An evaluation of the geologic evidence," *Geology* 10, (1982): 141–146.

[206] Ibid., 19, quoting Kenneth M. Towe, "Environmental Oxygen Conditions During the Origen and Early Evolution of Life," *Advances in Space Research* 18, (1996): (12)7–(12)15.

[207] Ibid., 20, quoting Philip H. Abelson, "Chemical Events on the Primitive Earth," *Proceedings of the National Academy of Sciences USA* 55, (1966): 1365–1372. (Emphasis original.)

[208] Ibid., 21, citing Gordon Schlesinger and Stanley L. Miller, "Prebiotic Synthesis in Atmospheres Containing CH4, CO, and CO2: I. Amino Acids," *Journal of Molecular Evolution* 19, (1983): 376–382.

- The accumulation of hydrogen in the Miller-Urey apparatus (up to 76 percent) has also been criticized since free hydrogen on the early earth would have escaped into space because of its light weight. Wells reports that by 1995, a near-consensus of geochemists and origin of life researchers had dismissed the Miller-Urey experiment because "the early atmosphere looked nothing like the Miller-Urey simulation."[209]
- One origin of life researcher especially critical of the Miller-Urey experiment has written, "We have reached a situation where a theory has been accepted as fact by some, and possible contrary evidence is shunted aside"—a situation he called "mythology rather than science."[210] One science writer for the *New York Times* has admitted, "Everything about the origin of life on earth is a mystery, and it seems the more that is known, the more acute the puzzles get."[211]

So was a "near-consensus" dismissal of the experiment by origin of life researchers enough to disqualify the Miller-Urey experiment as a favorite icon of evolution? Far from it. As noted above, Wells reviewed ten biology textbooks published between 1998 and 2000— all of which discussed the Miller-Urey experiment. Four of the ten received a grade of "D" for their treatment of this icon. The other six received an "F."

2. *Darwin's Tree of Life*

The central point of Jonathan Wells' discussion of the "tree of life" cited nonstop by advocates of Darwin's big idea is that the actual fossil record looks nothing like the picture of countless branches and

[209] Ibid., quoting Jon Cohen, "Novel Center Seeks to Add Spark Origins of Life," *Science* 270, (1995): 1925–1926.
[210] Ibid., 27, quoting Robert Shapiro, *Origins: A Skeptics Guide to the Creation of Life on Earth* (New York: Summit Books, 1986), 112.
[211] Ibid., 24, quoting Nicholas Wade, "Life's Origins Get Murkier and Messier," *The New York Times* (Tuesday, June 13, 2000), D1-D2.

twigs stemming from a single trunk. The true picture is more like a "lawn of life" where each of the major groups of creatures actually appeared suddenly, at about the same time, in near final form, without any transitional ancestors. See the discussion of the tree of life above in chapter 2.

3. *Homology in Vertebrate Limbs*

Homologous structures are ones that exhibit a very similar pattern of structure in very different species. The forelimbs of animals with backbones (vertebrates) is a classic example. It turns out that the bone structure of a man's arm, a dog's front leg, a bird's wing, and a whale's front flipper all exhibit a similar pattern, as shown by the various colored bones in figure 17. These are just four examples. The forelimbs of many other vertebrates show a similar pattern.

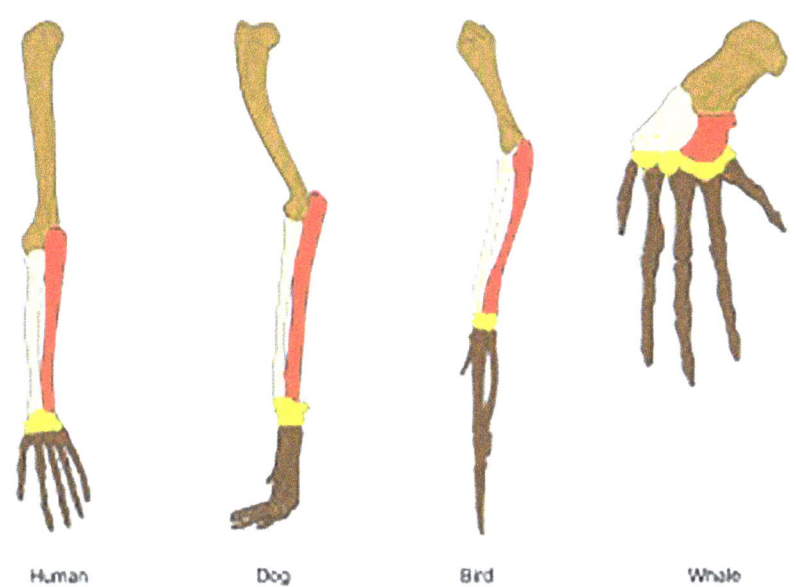

Human   Dog   Bird   Whale

Figure 17. Each of the species has an upper arm humerus, a forearm radius and ulna, and carpal and metacarpal bones plus digits in the hand, foot, wing, or fin.

Darwin argued that such homology was strong evidence that all vertebrates shared a common ancestor who had contributed some common structure that had been modified for each species by natural selection.

> The framework of bones being the same in the hand of a man, wing of a bat, fin of the porpoise, and leg of the horse…at once explain themselves on the theory of descent with slow and slight successive modifications.[212]

Jonathan Wells defines a major problem with using homologous bone structures as evidence of descent from a common ancestor by describing the error of a professor of evolution at Ohio State University named Tim Berra. Professor Berra wrote a book for the purpose of defending the theory of evolution against criticisms from creationists. As many evolutionists do, he was referring to the fossil record of homologous bone structures as strong evidence for Darwin's idea of common descent with modification over time. And he drew an analogy to automobile design:

> If you compare a 1953 and a 1954 Corvette, side by side, then a 1954 and a 1955 model, and so on, the descent with modification is over-whelmingly obvious. This is what [paleontologists] do with fossils, *and the evidence is so solid and comprehensive that it cannot be denied by reasonable people.*[213]

But, of course, the meaning of "descent" concerning a series of car models bears no relationship to the meaning of descent concerning the relationship between one generation of an organism and the

---

[212] Darwin, *Origin,* 451.
[213] Tim Berra, *Evolution and the Myth of Creationism* (Stanford, CA: Stanford University Press, 1990), 117–119. (Emphasis original.)

next. Darwin's idea of descent with modification by random muta-tion and natural selection imagines an entirely random process with no intended outcome in mind. Contrast this with the process by which cars are intentionally designed by a team of intelligent design-ers intending to preserve and improve on a successful brand image from one model year to the next.

So the similarity of design between two living species could either mean they had the same ancestors or it could mean they had the same designer—but the similarity of design itself does not even begin to tell us which one it is.

Since 1859, evolutionists have apparently offered only two pos-sible ideas for proving that the homology of vertebrate forelimbs is the result of something contributed by a common ancestor. Jonathan Wells shows that both of these ideas have been disproven:

- The first, exceedingly vague conjecture, was that a com-mon ancestor had contributed some type of "developmen-tal pathway" by which the embryos of different species would produce similar structures in the mature adults. But Wells reports that embryologists have known this is not the case for more than one hundred years. In recent decades, scientists have recognized that the whole idea that homol-ogous limbs develop from a common ancestral pattern in embryos only arose because proponents of evolution were superimposing their preconceptions on the evidence.[214]
- The only other idea for evidence that homologous limbs were produced by common descent has been that vertebrates with homologous forelimbs share similar genes contributed by a common ancestor. But at least during the DNA-age now more than fifty years old, biologists have known that this is not the case either. In 1971, one famous evolution-ary embryologist wrote concerning the genetics of homol-ogy, "This is where the worst shock of all is encountered... [because] characters controlled by identical genes are not

---

[214] Wells, *Icons*, 71–73.

necessarily homologous…[and] homologous structures need not be controlled by identical genes…[T]he inheritance of homologous structures from a common ancestor… cannot be ascribed to identity of genes."[215]

Despite the complete lack of evidence establishing that homologous forelimbs in vertebrates are due to common descent rather than a common designer, this icon of evolution is still relied on heavily in biology texts as evidence for common ancestry of species, such as bats, whales, and humans. Of the ten biology textbooks graded by Wells, seven received a "D" and three received an "F." A grade of "D" meant that homology was attributed either to similar genes or similar developmental pathways but failed to mention that the evidence supports neither. A grade of "F" meant the book also defined the term "homology" to mean similarity because of common ancestry before citing homology as evidence for common ancestry—in other words, reasoning in a perfect circle.

4. *Haeckel's Embryos*

You probably recall that in the preface, I recounted that my college anthropology course taught me the idea that embryo development provides strong evidence for the truth of macroevolution. Darwin himself thought this embryonic evidence was "by far the strongest single class of facts in favor of" his theory of macroevolution.[216] In his later book *The Descent of Man* (1871), Darwin reemphasized his idea in the context of humans:

> The [human] embryo itself at a very early period
> can hardly be distinguished from that of other

---

[215] Gavin de Beer, *Homology: An Unsolved Problem* (London, Oxford University Press, 1971), 15–16.

[216] September 10, 1860, letter from Darwin to Asa Gray, in Francis Darwin (editor), *The Life and Letters of Charles Darwin* (New York: D. Appleton & Company, 1896), Vol. II, 131.

members of the vertebrate kingdom. [Since humans and other vertebrates] pass through the same early stages of development…we ought frankly to admit their community of descent.[217]

But Darwin was not an embryologist, so his thought about the power of the embryological evidence depended entirely on the quality of the work of the embryologists he consulted. One of those embryologists was a German professor named Ernst Haeckel who was famous for his series of drawings of embryos and his contentions about how their similar form provided evidence of common descent. Darwin said that Haeckel "brought his great knowledge and abilities to bear on what he calls phylogeny, or the lines of descent of all organic beings. In drawing up the several series he trusts chiefly to embryological characters."[218] To summarize his "biogenetic law," Haeckel actually coined the phrase taught to me and countless others: "ontogeny recapitulates phylogeny."[219] Roughly, this means "development recounts descent."

The major problem with all this, as Wells describes it, is that "biologists have known for over a century that Haeckel *faked* his drawings; vertebrate embryos never look as similar as he made them out to be."[220]

Haeckel's drawings of certain vertebrate embryos are shown in figure 18. These drawings suggest that all vertebrate embryos start out looking almost identical at the beginning of their development which later, in effect, retraces the evolutionary adaptations of various ancestors as depicted in the "tree of life."

---

[217] Wells, *Icons*, 82.
[218] Ibid.
[219] Ibid., 87.
[220] Ibid., 82. (Emphasis original.)

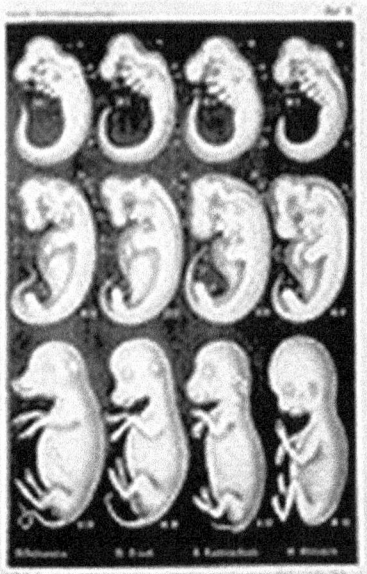

Figure 18. Drawings from Haeckel's 1874 book Anthropogenie. Stages from top to bottom called "very early," "somewhat later," and "still later." Left to right showing a fish, salamander, turtle, chick, pig, cow, rabbit, and human.

Jonathan Wells details a series of problems with the vertebrate embryos drawn by Haeckel and relied on by Darwin.[221] Taken together, these points make it almost impossible to avoid the conclusion that Haeckel intentionally distorted the "evidence" in order to present embryologic support for Darwin's theory of common descent.

- Haeckel claimed that his drawings began with a depiction of embryos at a "very early" stage—the point at which Darwin said such similarities were most important—but, in fact, the drawings actually *omit the earliest stage entirely* and begin at a point closer to midway through the embryos' development when their form is more similar than at any other stage. Even as of the late 1800s, embryologists knew that the embryos of different species were "*distinct and*

---

[221] Wells, *Icons*, 90–101.

distinguishable from its allies from the very earliest stages all through development."[222]

- While Haeckel was claiming to show great similarity across the broad subphylum of vertebrates, that is not what he did. The subphylum of vertebrates actually includes classes called jawless fishes and cartilaginous fishes as well as bony fishes, amphibians, reptiles, birds, and mammals. But Haeckel left out the first two classes, probably because their embryos are most unlike the others.

- For the class of amphibians, Haeckel ignored the most common species, the frog, in favor of a much more rare species, the salamander, probably because at one point of development, the salamander embryo looks much more like those of other classes.

- Even though the class of mammals is just one of seven vertebrate classes, fully half of Haeckel's drawings of vertebrate embryos were mammals—again, artificially increasing the appearance of similarity across the entire subphylum.

- Vertebrate embryos actually vary tremendously in size—from less than one millimeter to almost ten millimeters at the same stage of development—but Haeckel drew them all to be about the same size. Moreover, different species of vertebrate embryos actually have wide variation in the number of repetitive blocks of cells growing along their developing backbone (from eleven to more than sixty), but Haeckel's drawings show about the same number of blocks in each class.[223]

One of the most famous proponents of evolution, the late Harvard professor Stephen Jay Gould, has admitted many of the distortions noted by Wells. In an article titled "Atrocious!" written

[222] Adam Sedgwick, "On the Law of Development commonly known as von Baer's Law; and on the Significance of Ancestral Rudiments in Embryologic Development," *Quarterly Journal of Microscopical Science* 36, (1984): 35–52. (Emphasis original.)
[223] Wells, *Icons*, 90–99.

in the year 2000, Gould said that Haeckel "exaggerated the similarities by idealizations and omissions." He said scientists should all be "astonished and ashamed by the century of mindless recycling that has led to the persistence of these drawings in a large number, if not the majority, of modern textbooks." He concluded that Haeckel's drawings suffered from "inaccuracies and outright falsification" and were, in some cases, actually "fraudulent." In short, Gould referred to what had happened as *"the academic equivalent of murder."*[224]

Figure 19 shows embryos comparable to Haeckel's without the idealizations and omissions.

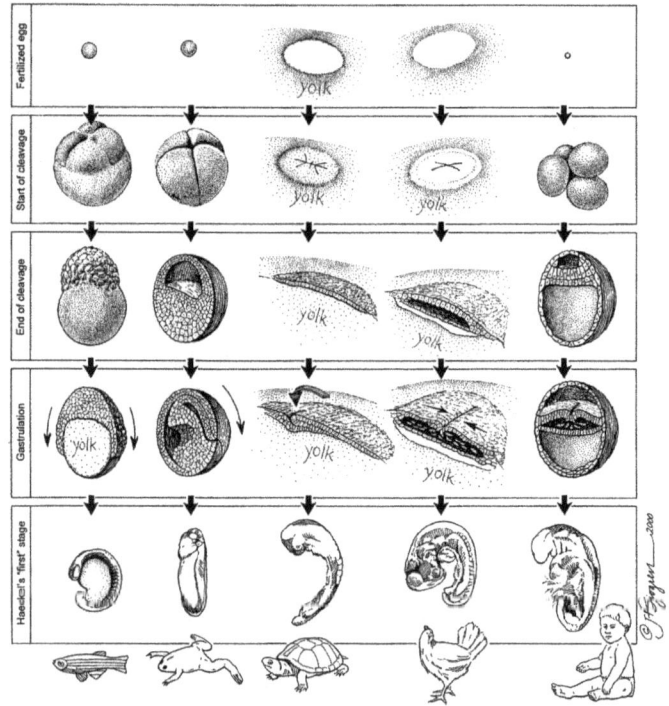

Figure 19. More accurate drawings of vertebrate embryos from the earliest stage. From left to right showing a fish, frog, turtle, chick, and human. (Copyright Jody F. Sjogren 2000. Used with permission.)

---

[224] Stephen Jay Gould, "Abscheulich! (Atrocious!)" Natural History (March, 2000), 42–49. (Emphasis added.)

As we will see again, apparently even being "the academic equiv-alent of murder" is not enough to disqualify an icon of evolution from continued use in many of the favorite high school and college biology textbooks—including a textbook written by the president of the National Academy of Sciences. "Like a zombie that just won't die, these bogus drawings keep coming back."[225] Of the ten text-books graded by Wells on how they treated Haeckel's embryos, two of the ten received a grade of "D" and the remaining eight all got an "F"—meaning that they use Haeckel's drawings without any men-tion of the many problems identified by Wells.

5. *Archaeopteryx: The Missing Link*

The central point of Jonathan Wells' discussion of *Archaeopteryx* is that while it is cited endlessly as a "missing link" between reptiles and birds, the evidence shows it to be an extinct dinosaur rather than an ancient relative of modern birds. See the discussion of the fossil record above in chapter 3, including figure 5.

6. *Peppered Moths*

Darwin closed the introduction of his most famous book by declaring that he was "convinced that Natural Selection has been the main but not exclusive means of modification" of species.[226] But he had no direct proof of natural selection ever having occurred. Instead, he was forced merely to "beg permission to give one or two imaginary illustrations."[227]

As the twentieth century dawned, some British scientists noticed that during the industrialization of Britain over the prior one hundred years, many members of a species known as peppered

---

[225] Casey Luskin, "Haeckel's embryo drawings make cameos in proposed Texas instructional materials," *Evolution News and Views* (June 17, 2011), http://www.evolutionnews.org/2011/06/haeckels_embryos_make_multiple047321.html.

[226] Darwin, *Origin*, 69.

[227] Ibid., 138.

moths had become darker in color. In the early 1800s, the moths had been quite light in color, but as industry had polluted certain major cities, the moths inhabiting those cities had become darker. This was known as "industrial melanism." By the year 1900, for example, more than 90 percent of the peppered moths found in the area of Manchester, England, were of the darker "melanic" variety. But the cause for this melanism remained unclear.

Figure 20. Melanic peppered moth above a typical moth below.

In the 1920s, a British biologist named J.W. Harrison reported that melanism could be produced in moths by feeding moth larvae leaves containing metallic salts—in other words, by feeding airborne industrial pollutants. But Harrison's work was criticized for a variety of reasons. As the theory of evolution became more and more widely accepted, biologists began to assume that darker moths were better camouflaged than lighter moths on trees that had been darkened by pollution. In other words, they presumed that Darwin's mechanism of natural selection explained the change.

But it was not until the 1950s that a British biologist named Bernard Kettlewell decided to put this theory to the test by conducting a series of experiments using peppered moths—beginning in the laboratory and later moving into the field. His field experiments involved marking two groups of moths—darker moths called "melanics" and lighter moths called "typicals"—with tiny dots of paint under their wings, and then releasing them onto tree trunks during the day. Kettlewell then set traps to recapture as many of his subjects as he could the following night.

In one trial where he released his moths onto tree trunks darkened by pollution, Kettlewell recovered 27.5 percent of the darker melanic moths but only 18 percent of the typicals. In a subsequent trial where moths were released onto lighter tree trunks, he recaptured about twice as many typicals as melanics—almost completely reversing the ratio of the prior trial.

The results seemed to be dramatic evidence of natural selection at work. Kettlewell published an article in which he claimed to present "Darwin's missing evidence" and called his results "the most striking evolutionary change ever witnessed in any organism."[228]

Though hailed as proof of Darwinian evolution—only microevolution, mind you, not macro—Kettlewell's results were actually worthless because of fundamental flaws in his experimental procedure:

- Peppered moths actually fly during the night, find a resting place near dawn, and remain stationary during the day. But Kettlewell faced too many practical difficulties to conduct his experiments in the dark. So when he began his experi-

---

[228] H.B.D. Kettlewell, "Darwin's Missing Evidence," *Scientific American* 200 (March 1959), 48–53.

ments during the day by releasing moths onto tree trunks, they just sat there as easy targets for predatory birds.[229]

- Even more importantly, *peppered moths do not normally rest on tree trunks at all.* Subsequent research has determined that their normal resting place, selected as dawn approaches, is on the bottom side of more or less horizontal branches high up in the canopy. This means that predatory birds would be much less likely to see them at all. One researcher reported that in twenty-five years of fieldwork, he had only ever seen *one* peppered moth naturally perched on a tree trunk.[230]

In addition to these procedural errors, the Kettlewell results have been contradicted by the evidence resulting from changes in tree color produced by environmental laws. Studies showed that melanic moths did well in regions with little pollution and typicals did well in some regions with conditions suggesting their camouflage would be poor.[231] Taken together, these criticisms led biologists to conclude "the evidence Darwin lacked, Kettlewell lacked as well."[232]

Nevertheless, Wells reports that "[a]lmost every textbook that deals with evolution not only re-tells the classical peppered moth story without mentioning its flaws, but also illustrates it with staged photographs."[233] By "staged," Wells means the book contains pictures of dead peppered moths either *glued or pinned to tree trunks.* Of the ten biology textbooks graded by Wells, nine of them presented this icon of evolution. Wells gave a grade of "D" to three of the nine and the grade of "F" to the other six.[234] The grade of "F" meant that the textbook told the story of peppered moths as evidence of Darwinian natural selection without mentioning any of the flaws in the classical story

---

[229] Wells, *Icons*, 148.

[230] Ibid., 149.

[231] Ibid., 144–148.

[232] Giuseppe Sermonti and Paola Catastini, "On industrial melanism: Kettlewell's missing evidence," *Rivista di Biologia* 77 (1984), 35–52.

[233] Wells, *Icons*, 156.

[234] Wells chose not to give any grade to one of the nine because while it told the familiar peppered moth story, it contained no picture of peppered moths.

and contained pictures of moths on tree trunks without mentioning that resting on tree trunks is not the natural habit of such moths.

7. Darwin's Finches

The central point of Jonathan Wells' discussion of Darwin's Galapagos finches is that what Darwin observed in those Pacific island birds was merely microevolution, not macroevolution. In addition, recent study of the genetics of those finches is an important part of the proof that macroevolution not only has not occurred but that it cannot occur. See the discussion of Darwin's finches and the microbiology of evolution in chapter 5.

8. Four-Winged Fruit Flies

Darwin's big idea was built on the presumption that natural selection preserved natural variations in the members of a species that could be transmitted from one generation to the next. But Darwin didn't know much of anything about genetics. Gregor Mendel was alive during Darwin's time, but his work on genetics was not really appreciated until after the turn of the twentieth century. Nor did Darwin know about the idea that a "gene" could "mutate"—two terms that did not even exist until the turn of the century.[235]

But as widespread acceptance of Darwinism and the study of genetics grew, genetic mutations came to be viewed as the source of the adaptive variations needed for macroevolution to work. As noted above, by 1937, a major work synthesizing the theory of evolution with the science of genetics declared, "Mutations and chromosomal changes…constantly and unremittingly supply the raw materials for evolution."[236]

But in order to provide "the raw materials for evolution," the beneficial mutation has to actually affect the shape of an organism—

---

[235] Wells, *Icons*, 180.

[236] Theodosius Dobzhansky, *Genetics and the Origin of Species* (New York: Columbia University Press, 1937), 13.

its "morphology"—because without such change, one species can never change in a way that eventually will become a different species. Such beneficial morphological mutations are exceedingly rare.[237]

So it is easy to imagine the excitement among biologists in 1978 when a geneticist at the California Institute of Technology reported breeding a mutant fruit fly that had four normal-looking wings rather than just two. Since 1978, pictures such as figure 21 have become an increasingly popular icon of evolution appearing in biology textbooks and many other public presentations of evolution.[238]

But there's a problem. Actually, there are two big problems—each of which destroys the value of this four-winged creature as any evidence for Darwin's big idea:

- First, the four-winged fruit fly was not a naturally occurring creature but the product of careful breeding by geneticists who were *combining three different strains of mutant flies that had been maintained artificially as laboratory stock for decades.* The first mutation, which had occurred in 1915, enlarged the tiny appendages behind the wings which the fly uses to balance properly during flight. This mutant variety had been preserved ever since. The geneticist in 1978 had bred these mutant flies with flies with a different mutation, making the "balancers" slightly bigger, and then bred the offspring of these two mutants with yet another different mutant to produce the triple-mutant fly depicted at the bottom of figure 21. Such a planned crossbreeding program would never have occurred in the wild.[239]
- Furthermore, the extra pair of wings on the four-winged fruit fly lacks flight muscles, *making these wings completely nonfunctional.* With this useless pair of wings hanging on its body, this fly is seriously disabled. "[B]ecause of this, four-winged males have difficulty mating, and unless the

[237] Wells, *Icons*, 181–82.
[238] Ibid.
[239] Ibid., 182–186.

line is carefully maintained in a laboratory it quickly dies out."[240]

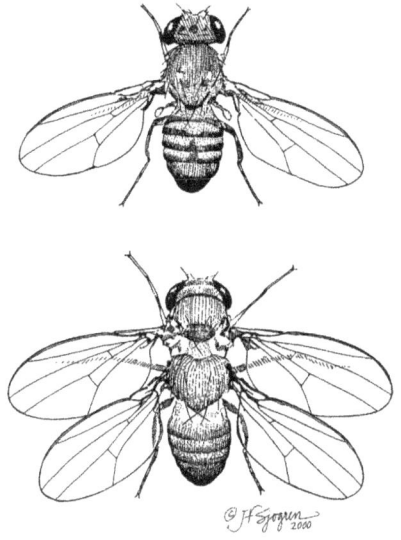

Figure 21. Above is a normal fruit fly with normal flight balancers just behind its wings. Below is a four-winged fly with balancers that have been transformed into normal-looking, but dysfunctional, wings. (Copyright Jody F. Sjogren 2000. Used with permission.)

So just as in 1859, the theory of evolution has plenty of evidence showing how small morphological changes might be preferred and preserved by natural selection, such as in the beaks of finches on the Galapagos Islands. But despite an extraordinary amount of work—including an army of researchers artificially creating mutations in countless generations of fruit flies for well more than one hundred years—Darwin's big idea still has no evidence of a morphological mutation producing a functional adaptive change of form

---

[240] Ibid., 186.

that could be seen as changing one kind of creature into an entirely different kind.

9. *Fossil Horses*

Many biologists have been quick to claim that the evolutionary lineage of the modern horse is better supported by the fossil evidence than the lineage of any other animal. Such claims are often accompanied by illustrations such as figure 22, showing a small four-toed ancient horse becoming a large single-toed modern horse over a period of fifty million years.

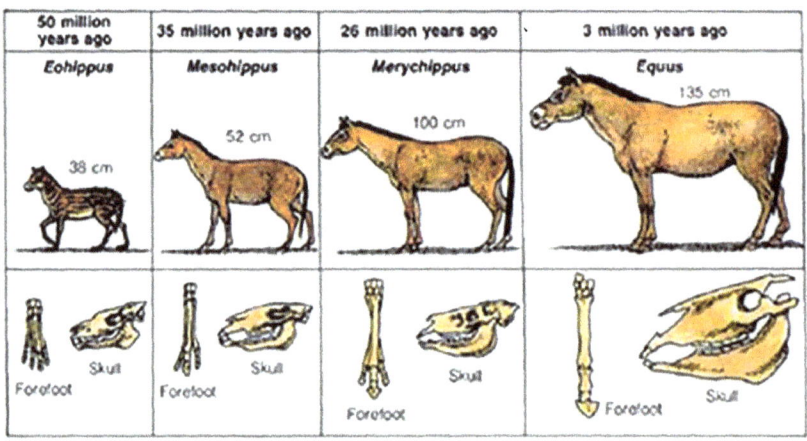

Figure 22. Illustration depicting fifty million years of horse evolution.

Jonathan Wells points out, however, that the actual fossil evidence is not so simple.

> Early versions of these illustrations showed horse evolution proceeding in a straight line from the primitive ancestor through a series of intermediates to the modern horse. But paleontologists soon learned that horse evolution was much more complicated than this. Instead of being a

linear progression from one form to another, it appeared to be a branching tree, with most of its branches ending in extinction.[241]

It seems to me, however, that the horse evolution icon does not deserve extended discussion for one simple fact illustrated by the reality of dog breeding. A three-pound Chihuahua and a 120 pound Great Dane look almost nothing alike. But they are both still dogs. The ancient four-toed horse and the modern single-toed horse also are somewhat different. But they are both still horses.

Figure 23. Chihuahua on the left; Great Dane on the right.

As we have seen repeatedly, there is good evidence supporting the idea of microevolution but no convincing evidence supporting the idea of macroevolution. The icon of horse evolution may support the idea of microevolution but not Darwin's big idea of macroevolution.

[241] Wells, *Icons*, 195.

10. *From Ape to Human: The Ultimate Icon*

Jonathan Wells calls it the ultimate icon of evolution—so important that he used it to illustrate the cover of his book.

Figure 24. The Ultimate Icon. (Copyright Regnery Publishing, graphic design, and Jody F. Sjogren, illustration, 2000. Used with permission.)

I'm sure you've seen an illustration similar to this before now. It shows a series of creatures from a smaller knuckle-dragging ape through a series of bipedal forms up to a modern man. Almost never does the illustration tell the reader that the creatures depicted are *entirely hypothetical*—not reconstructions based on any fossil evidence at all. They exist only in the mind of the artist and now in many other minds as well. Well-known Harvard evolutionist Stephen Jay Gould admitted that this icon "is *the* canonical representation of evolution—the one picture immediately grasped and viscerally understood by all."[242]

Viscerally understood but undeniably false. In fact, there is no fossil evidence supporting the parade of creatures shown in this ultimate icon. One science journalist has reported that the scientific evidence relied on by paleoanthropologists in an effort to reconstruct man's evolutionary history is

> a pitifully small array of bones…One anthropologist has compared the task to that of reconstructing the plot of *War and Peace* with 13 randomly selected pages.[243]

Indeed, the chief science writer for the scientific journal *Nature* once admitted that all the physical evidence for human evolution

> between about 10 and 5 million years ago—several thousand generations of living creatures—can be fitted into a small box.[244]

Other well-known evolutionists have admitted that if the common claim were true that fossils explain the evolutionary ancestry of humans,

[242] Stephen Jay Gould, *Wonderful Life: The Burgess Shale and the Nature of History* (New York–London: W.W. Norton & Company, 1989), 31. (Emphasis original.)

[243] Constance Holden, "The Politics of Paleoanthropology," *Science*, 213 (1981), 737–740.

[244] Henry Gee, *In Search of Deep Time: Beyond the Fossil Record to a New History of Life* (New York: The Free Press, 1999), 202.

one could confidently expect that as more homi-
nid fossils were found the story of human evolution would become clearer. Whereas if anything,
the opposite has occurred.[245]

Yet another leading British primate expert Solly Zuckerman has
pointed to its claims on human origins to question the legitimacy of
the entire field of anthropology.

Zuckerman's judgment of the professional stan-
dards of physical anthropology was not a gener-
ous one: he compared it to parapsychology and
remarked that the record of reckless speculation
in human origins "is so astonishing that it is
legitimate to ask whether much science is yet to
be found in this field at all."[246]

Nevertheless, as usual, the advocates of evolution struggle to
hang on to their favorite theory despite the facts. So even the ultimate icon of evolution lives on, falsely suggesting to children and
adults alike that we are all nothing more than modern monkeys.

---

[245] Niles Eldredge and Ian Tattersall, *The Myths of Human Evolution* (New York:
Columbia University Press, 1982), 126–127.
[246] Johnson, *Darwin On Trial*, 84.

CONCLUSION: DARWIN'S BIG IDEA IS FABLE, NOT FACT

Why Does the Evidence Never Fit the Theory?

So if the historical record of life on earth does not show a gradual progression from simple life to more complex, but instead shows both plants and animals exploding onto the scene all at once without partially formed ancestors, then how can Darwin's big idea be true?

If the infinite number of transitional forms predicted by Darwin don't exist but, at most, his advocates can scrape together only a handful of controversial transitional candidates, then how can Darwin's big idea be true?

If after more than 150 years of scholarship no evolutionist has ever even attempted to explain specifically how any complex organ or biological system could have evolved gradually by means of mutation and natural selection, then how can Darwin's big idea be true?

If the only studies at the genetic level show that adaptive mutations have been the result of the damage or destruction of genetic information, yet we know that new information is required for the emergence of more complex body plans and systems, then how can Darwin's big idea be true?

If it would take a hundred million or a billion generations of mutations for only a two-step mutational improvement to occur but biology is full of intricate systems that would require at least dozens, if not hundreds, of such steps, then how can Darwin's big idea be true on a planet where, at most, life has existed for only about four billion years?

If natural selection is supposed to preserve only traits that promote the survival or reproduction of organisms, why is nature full of traits that do neither?

If there is not even a single "icon of evolution" that is based on well-established physical evidence, then how can Darwin's big idea be true?

The answer is that *Darwin's big idea can't be true*. It never has been true, but now we know for sure.

In 1945, a biochemist named Ernst Chain was one of three winners of the Nobel Prize in medicine for his groundbreaking work in the mass production of one of the world's first groups of antibiotics called penicillin. As a senior scientist in 1971, Professor Chain expressed grave misgivings about Darwin's big idea.

> [Darwinian macroevolution] is a hypothesis based on no evidence and irreconcilable with the facts... This hypothesis willfully neglects the principle of teleological purpose [the appearance of design] which stares the biologist in the face wherever he looks...These classical evolutionary theories are a gross oversimplification of an immensely complex and intricate mass of facts, and it amazes me that they were swallowed so uncritically and readily, and for such a long time, by so many scientists without a murmur of protest.[247]

Ernst Chain wrote these words in 1971, the year I graduated from high school and began college at the University of Wisconsin. Unfortunately, *at no point during the many college lectures I heard concerning the theory of evolution did anyone ever mention that the theory ever had been criticized by anyone*, much less by a Nobel Laureate. That kind of thing was a very well-kept secret.

But now you know.

Since Ernst Chain wrote about his misgivings in 1971, the scientific evidence has grown by leaps and bounds—especially in the

---

[247] Wells, *Zombie Science*, 152–153, citing Ernst Chain, "Social responsibility and the scientist in modern western society," *Perspectives in Biology and Medicine* 14:3 (1971), 347–369.

field of genetics. And the more we have learned, the more problems have been identified concerning Darwin's big idea. As we have seen, the problems are insurmountable.

The Darwinian Paradigm Is Crumbling

Science is supposed to be tentative and self-correcting. Yet advocates of Darwin's big idea have been ignoring or rejecting contrary evidence for decades to protect their beloved theory.

Nevertheless, Darwinian skeptics are beginning to say with some confidence that the paradigm is beginning to crumble in the face of a growing mountain of contrary evidence—especially from the new and growing field of evolutionary developmental biology (often known as "evo-devo"). For example, the skeptic with whom I began in 1994, Michael Denton, has recently declared, "But despite these dyed-in-the-wool Darwinists, *there is now a growing chorus of dissent within mainstream evolutionary biology!*"[248]

One of the big reasons for this growing dissent is the accumulation of genetic research disproving well-settled predictions of modern Darwinism. For example, before genetic research enabled the genomes of creatures to be fully mapped, it was widely assumed by mid-twentieth-century Darwinists that

> over the hundreds of millions of years that have elapsed since the last common ancestor of the animal phyla, every gene would have been crafted and re-crafted multiple times...so that all evidence of any commonalities would have been lost millennia ago. Based on this widely held Darwinian expectation, Ernst Mayr, one of the giants of twentieth century evolutionary biology...claimed that "the search for homologous genes is quite futile except in closely related

---

[248] Denton, *Evolution: Still a Theory in Crisis*, 26. (Emphasis original.)

species." *There is hardly a prediction in science that has proved so woefully wrong!*[249]

The fact is that evo-devo research has now shown, for example, that a mutation in the very same gene causes maldevelopment in the eyes of both fruit flies and humans. Similarly, experimental work has identified another gene that plays the very same role in the development of shark fins as in the development of digits in both chickens and mammals.[250] The idea that the genetics of fruit fly eyes stayed the same while the genetics of everything else about the fruit fly changed enough to turn that fly into a human seems unthinkable.

At the same time, other well-accepted stories about the theory of macroevolution have been effectively disproven by recent research at the molecular level. It has long been imagined, for example, that legless fish somehow evolved into four-legged (tetrapod) amphibians. But modern research focusing on the actual genetics of limb development shows that the bony elements of fins are not ancestral to the legs or hands of the amphibian.[251]

> No matter what Darwinian evolutionary "spin" is put on the gap between fin and hand, there is no avoiding the fact that a significant break does exist in the natural order, and the new evo-devo picture provides no support for any sort of gradualist, functionalist scenario.[252]

The very same thing is true concerning the story of reptilian scales becoming avian feathers. For many decades, Darwinists have argued that scales somehow evolved into feathers as reptiles either jumped out of trees attempting to glide or ran after prey attempt-

---

[249] Ibid., 86. (Emphasis original.)
[250] Ibid., 83–84.
[251] Ibid., 160.
[252] Ibid., 165.

ing to fly. This was known as the "frayed reptile scale" paradigm of feather evolution. But the latest genetic research shows otherwise.

> Feathers, however, are hierarchically complex assemblages of numerous evolutionary novelties—the feather follicle, tubular feather germ, feather branched structure, interacting differentiated barbules—*that have no homolog in any antecedent structures.*[253]

> Neo-Darwinian approaches to the origin of feathers have failed to appropriately recognize the novel features of feather development and morphology, and have thus failed to adequately explain their origins.[254]

So as we have seen many times before, the closer one looks at the exact details concerning how Darwin's big idea might have produced the things we see in nature, the more one is forced to conclude these things did not occur as Darwin imagined.

Advocates of Darwin's big idea have long contended that the idea must be true since virtually all scientists agree with them, but that claim is getting harder and harder to make. As of May 2021, more than one thousand PhD scientists or academic MDs had signed their agreement with the following statement:

> We are skeptical of claims for the ability of random mutation and natural selection to account for the complexity of life. Careful examination

---

[253] Ibid., 172. (Emphasis original.)

[254] Ibid., 179, citing Richard O. Prum and Alan H. Brush, "The Evolutionary Origin and Diversification of Feathers," *The Quarterly Review of Biology* 77, no. 3 (September 2002), 289.

of the evidence for Darwinian theory should be encouraged.[255]

The scientists in agreement with this statement are associated with some of the most prestigious universities in the world, including Harvard, Yale, Princeton, MIT, Johns Hopkins, the University of Chicago, UC Berkeley, UCLA, Oxford, Cambridge, and many, many others. More than one thousand signatures is actually a remarkably large number considering the fact that academic employment and research funding are often denied to scientists who have challenged Darwin's big idea in any way.[256] For these reasons, it is extremely likely that more than one thousand is a *vast understatement* of the number of scientists who actually agree with this statement. So yes, scientific support for Darwin's big idea is beginning to crumble.

So If Not Macroevolution, Then What?

It's a good thing to discover that something often described as a well-proven fact is actually untrue. But it's not the best thing. The best thing is to know the truth. So what is the truth about the origin of all living things?

Years ago, when I had studied enough to realize that Darwin's big idea was contradicted by the evidence, I was truly shaken. I was a forty-year-old man with infant twin daughters who needed to be taught about everything in life, but their dad didn't even know something as basic as where people came from.

But I knew enough to know this: people either had evolved from something else or they must have been created. These seemed to be the only two possibilities.

As a forty-year-old, I was a nominal "Christian" by default. My father had taken me to a Lutheran church a handful of times as a

---

[255] https://dissentfromdarwin.org/
[256] https://www.youtube.com/watch?v=7VDVZ19wy5Q

child, and I had been given a Bible, but I had never read it. I knew nothing about God. In my twenty years of marriage, before learning the truth about evolution, my wife and I had never been to church even once. I was pretty sure that the Bible said something about creation, but I wasn't sure what that was.

Imagine my surprise when I read what Genesis had to say about creation and found out that it had more scientific support than Darwin's big idea! Here's what I mean:

- Genesis 1:1 begins the Bible with the words, "In the beginning, God created the heaven and the earth." And the New Testament refers multiple times to things that happened "before the foundation of the world."[257] But prior to the 1900s, most people thought that the universe was eternal—that time had *no beginning*. Indeed, it was not until April 1992 that scientists proved that matter, energy, the three dimensions of space, and time itself actually all began at the hot "big bang" creation event.[258] So how did Moses know at least three thousand years earlier that the creation event described in Genesis truly was the "beginning"? And how did the apostle Paul know so many centuries ago that the world actually had a "foundation" (or "founding")?

- The first chapter of Genesis describes eleven distinct creation events concerning the universe, the sun, the moon, the earth and its atmosphere, in addition to the appearance of plants and animals on earth. The time order in which these events are described "perfectly accords with the findings of modern science."[259] The odds of describing eleven events in perfect order just by chance is about one in forty million.[260] So how did Moses get all that right?

---

[257] See Ephesians 1:4.

[258] Hugh Ross, *The Creator and the Cosmos* (Colorado Springs, Colorado: NAVPRESS, 1993), 19–20.

[259] Hugh Ross, *The Fingerprint of God* (Orange, California: Promise Publishing Co., 1991), 168.

[260] Ibid., 169.

- Genesis says that God created the grasses and plants as well as "every living creature that moveth."[261] And concerning each and every category of plant and animal, Genesis declares that the plants and animals were brought forth "after his own kind."[262] So consistent with the fossil record, Genesis actually says that life-forms had not changed from one category to another but that each had and would reproduce after its own kind.

Learning that the Bible accurately described scientific facts unknown to humans thousands of years ago had a great impact on me. There had to be some explanation for that accuracy other than chance. This impact grew even stronger as I learned more and more about scientific truths described in the Bible many centuries before such facts were known to man. Here are some examples:[263]

- Ancient civilizations imagined that the earth may have been carried on the back of a gigantic turtle or held up by a mythological figure such as Atlas, a giant among giants. But about 3,500 years ago, the book of Job declared that God "stretcheth out the north over the empty place, and *hangeth the earth upon nothing*."[264] Human understanding of the earth as a globe hanging in space did not begin to be widely accepted until the theory of gravity was formulated by Sir Isaac Newton in the seventeenth century—and was not proven until Einstein's general theory of relativity in 1915. So how did the author of Job know about this in 1500 BC?
- The idea of "ocean currents," such as the Gulf Stream, is a familiar concept to us these days, but their discovery was

---

[261] Genesis 1:21, which specifically reports that "God created great whales."

[262] Genesis 1:11, 12, 21, 24, and 25.

[263] These few examples are a small sample of those presented in a little book by Ray Comfort called *Scientific Facts in the Bible: 100 Reasons to Believe that the Bible Is Supernatural in Origin* (Newberry, Florida: Bridge-Logos, 2001).

[264] Job 26:7. (Emphasis added.)

not that long ago. Matthew Maury (1806–1873) is considered the father of oceanography. After he discovered and mapped the ocean currents, he was nicknamed the "Pathfinder of the Seas." Mr. Maury actually went searching for ocean currents because the Bible told him they were there, and he believed it. The book of Psalms described "The fowl of the air, and the fish of the sea, and whatsoever passeth through *the paths of the seas.*"[265] So how did the author of Psalm 8 know that there were paths of the seas about one thousand years before they were discovered?

- Similarly, the use of the term "trade winds" dates back only to about the fourteenth or fifteenth century as the experience of ocean-going tradesmen taught them that there are permanent prevailing winds that blow in certain directions around the globe. But something like two thousand years earlier, the book of Ecclesiastes described, "The wind goeth toward the south, and turneth about unto the north, it whirleth about continually, and the wind returneth again according to his circuits."[266] How did the author of Ecclesiastes know that?

- We think of quarantines and sanitation to be the stuff of relatively modern medicine. Indeed, in the mid-1800s, untold numbers of patients were still dying in hospitals because when moving from one patient to the next, physicians were either not washing their hands at all or washing in bowls of stagnant water rather than in running water. Yet about three thousand years earlier, the book of Leviticus had set forth the following instructions for health: "And when he that hath an issue is cleansed of his issue; then he shall number to himself seven days for his cleansing, and wash his clothes, and bathe his flesh in running water, and shall be clean."[267] So how did the author of Leviticus

---

[265] Psalm 8:8. (Emphasis added.)
[266] Ecclesiastes 1:6.
[267] Leviticus 15:13.

understand that quarantines and running water were valuable for health about three thousand years before medical science learned these things?

- The Bible actually contains many health instructions based on knowledge of bacteriology written thousands of years before the work of people like Louis Pasteur. For example, the book of Exodus warned, "[N]either shall ye eat any flesh that is torn of beasts in the field; ye shall cast it to the dogs."[268] And again, "All the days wherein the plague shall be in him he shall be defiled; he is unclean: he shall dwell alone; without the camp shall his habitation be."[269]

- Bloodletting was performed in the hope of achieving therapeutic purposes for about two thousand years, from the ancient Egyptians through the early history of the United States. In fact, almost a gallon of blood was drained from president George Washington during the ten-hour period just prior to his death from a throat infection in 1799. Yet more than three thousand years earlier, the author of Leviticus had written a medical truth that might have been read by physicians as an instruction against bloodletting for the infirm. As part of a commandment not to eat blood, this important observation read, "For the life of the flesh is in the blood."[270]

- Prior to the early 1900s, scientists thought that the universe consisted entirely of the Milky Way Galaxy. But then the observations of Edwin Hubble proved both that the universe contained entirely separate galaxies outside of our own, and that the universe had been expanding at increasingly higher rates of speed ever since its beginning. This work proved the truth of the biblical phrase, repeated

---

[268] Exodus 22:31.
[269] Leviticus 13:46.
[270] Leviticus 17:11.

several times, that God "stretchest out the heavens like a curtain."[271]

- The Bible contains statements anticipating both the first and second laws of thermodynamics many centuries before they were defined in the 1800s. The first law of thermodynamics states that the form of both matter and energy can be changed, but that neither matter nor energy can be created or destroyed. Genesis 2:1 said essentially the same thing by summarizing the creation events of chapter 1 with the words: "Thus the heavens and the earth were finished, and all the host of them." The original Hebrew word translated "finished" meant that which had happened in the past would never again occur.[272] The second law of thermodynamics states that, over time, every ordered process tends to become more and more disordered. In other words, things wear out.[273] The Bible states repeatedly that the earth is wearing out. For example, "Lift up your eyes to the heavens, and look upon the earth beneath: for the heavens shall vanish away like smoke, and the earth shall wax old like a garment..."[274] and "Of old hast thou laid the foundation of the earth: and the heavens are the work of thy hands. They shall perish, but thou shall endure: yea, all of them shall wax old like a garment; as a vesture shalt thou change them, and they shall be changed."[275]

And so I asked myself concerning each and every Bible statement of modern scientific fact, "How can this be?" If Moses was the author of the first books of the Old Testament, how could it be that as the author of Genesis or Exodus or Leviticus (written about 3,500 years ago), Moses could have described a series of truths about physics

---

[271] Psalm 104:2.
[272] Comfort, *Scientific Facts in the Bible*, 14.
[273] Ibid., 13.
[274] Isaiah 51:6.
[275] Psalm 102:25–26, which is repeated at Hebrews 1:11.

or biology or oceanography only discovered in the last few hundred years? Or how could the author of Isaiah or Job or the Psalms do exactly the same? By guesswork? And hit the mark every time?

No, there had to be some other explanation.

Christians I talked to had another explanation—an explanation that comes from the Bible itself. The explanation was that Moses and the other human authors were not working alone but were writing what God told them to write. This meant that the Bible was able to report true facts about the universe and the earth because it was authored by the Creator of the universe and the earth. The Bible reported true facts about biology and oceanography and physics because it was authored by the One who had created all those things as well. The idea was that the Bible contains words inspired by God Himself.[276]

I considered these claims just like I consider all other claims, by focusing on one central question: "Does the claim square with the evidence?" After months of careful consideration, I concluded that it did.[277]

---

[276] The explanation is repeated in both the Old Testament and in the New Testament. In the Old, the Bible reminded Hebrew readers of God's delivery to them of the Ten Commandments and His delivery of them out of their captivity in Egypt and that He fed them "manna" in the wilderness "which thou knewest not, neither did thy fathers know; that he might make thee know that man doth not live by bread only, but by every word that proceedeth out of the mouth of the Lord doth man live" (Deuteronomy 8:3). In the New Testament, the apostle Paul told his helper Timothy, "All scripture is given by inspiration of God" (2 Timothy 3:16).

[277] There are many fine reasons for trusting the truth of the Bible as the inspired Word of God (in the original languages). See Trinitarian Bible Society, Article 116, and *The Divine Inspiration of the Holy Scriptures* (London, England, 1987, 1999), 19–20. (Here is a book which has been before the world for more than 1,800 years, the busiest and most changeful period the world has ever seen… But all this time, men have never discovered a weak point or a defect in the Bible. Infidels have assailed it in vain… The march of intellect never overtakes it. The wisdom of men never gets beyond it. The science of philosophers never proves it wrong. The discoveries of travelers never convict it of mistakes… There is only one account to be given of the fact—the Bible was written by inspiration of God.)

My conclusion that Darwin's big idea is a fraud did not happen because I believed the Bible was true. For me, faith in the Word of God actually came much later. But both things happened because I followed the evidence to its logical conclusion—something you also are now well-equipped to do.

Now It's Your Turn

So now you know some of the best-kept secrets about evolution. And hopefully, now you also understand how to find your way to the truth. Please tell your children and your parents. Tell your brothers and sisters. Tell some friends, especially those with small children. Help them all realize that the popular story that people are simply a mindless accident of nature is nothing but a blatant lie.

BIBLIOGRAPHY

Behe, Michael J., *Darwin's Black Box: The Biochemical Challenge to Evolution* (New York, NY: The Free Press, 1996).

_____, *Darwin Devolves: The New Science About DNA That Challenges Evolution* (New York, NY: HarperCollins, 2019).

Bell, Graham, *The Masterpiece of Nature: The Evolution and Genetics of Sexuality* (Berkeley, CA: University of California Press).

Berra, Tim. *Evolution and the Myth of Creationism* (Stanford, CA: Stanford University Press, 1990).

Brockman, John, editor, *The Third Culture: Beyond the Scientific Revolution* (New York: Simon & Schuster, 1995).

Comfort, Ray, *Scientific Facts in the Bible: 100 Reasons to Believe that the Bible Is Supernatural in Origin* (Newberry, Florida: Bridge-Logos, 2001).

Darwin, Charles, *Origin of Species* (New York: Random House, 1979, originally published in 1859 by J. Murray, London, under the title, *On the Origin of Species by Means of Natural Selection*).

Dawkins, Richard, *River Out of Eden* (New York: Basic Books, 1995).

_____, *The Blind Watchmaker: Why the Evidence of Evolution Reveals a Universe Without Design* (New York, NY: W.W. Norton & Co., 1996).

de Beer, Gavin, *Homology: An Unsolved Problem* (London, Oxford University Press, 1971).

Denton, Michael, *Evolution: A Theory in Crisis, New Developments in Science are Challenging Orthodox Darwinism* (Bethesda, MD: Adler & Adler, 1985).

_____, *Evolution: Still a Theory in Crisis* (Seattle, WA: Discovery Institute Press, 2016).

Dobzhansky, Theodosius, *Genetics and the Origin of Species* (New York: Columbia University Press, 1937).

Eldredge, Niles and Tattersall, Ian, *The Myths of Human Evolution* (New York: Columbia University Press, 1982).

Fodor, Jerry and Piattelli-Palmarini, Massimo, *What Darwin Got Wrong* (London: Profile Books Ltd., 2010).

Gee, Henry, *In Search of Deep Time: Beyond the Fossil Record to a New History of Life* (New York: The Free Press, 1999).

Gould, S. J., *The Panda's Thumb* (New York and London: W.W. Norton and Co., 1980).

_____, Luria, S.E., and Singer, S., *A View of Life* (Menlo Park, CA: The Benjamin/Cummings Publishing Company, Inc., 1981).

_____, *Wonderful Life: The Burgess Shale and the Nature of History* (New York–London: W.W. Norton & Company, 1989).

Hawking, S.W., *A Brief History of Time: From the Big Bang to Black Holes* (New York: Bantam Books, 1988).

Johnson, Phillip, *Darwin on Trial* (Downers Grove, Illinois: Intervarsity Press, 1991).

_____, *Defeating Darwinism by Opening Minds* (Downers Grove, Illinois: Intervarsity Press, 1997).

_____, *The Wedge of Truth: Splitting the Foundations of Naturalism* (Madison, WI: InterVarsity Press, 2000).

Maddox, John, *What Remains to be Discovered* (New York: The Free Press, 1998).

Montefiore, Bishop, *The Probability of God* (London: SCM Press 1985).

Prothero, Donald R., *Evolution: What the Fossils Say and Why It Matters* (New York: Columbia University Press, 2007).

Remine, Walter James, *The Biotic Message: Evolution versus Message Theory* (St. Paul, MN: St. Paul Science, 1993).

Ridley, Mark, *The Cooperative Gene* (New York: The Free Press, 2001).

Ross, Hugh, *The Fingerprint of God* (Orange, California: Promise Publishing Co., 1991).

_____, *The Creator and the Cosmos* (Colorado Springs, Colorado: NAVPRESS, 1993).

Shapiro, Robert, *Origins: A Skeptics Guide to the Creation of Life on Earth* (New York: Summit Books, 1986).

Stanley, S., *Macroevolution* (San Francisco: W.H. Freeman and Co., 1979).

Wells, Jonathan, *Icons of Evolution: Science or Myth?* (Washington, DC: Regenery Publishing, 2000).

_____, *Zombie Science: More Icons of Evolution* (Seattle, WA: Discovery Institute Press, 2017).

ABOUT THE AUTHOR

Bob Kirk was born in Chicago and grew up on a farm in a tiny town in Northern Wisconsin. He graduated Phi Beta Kappa from the University of Wisconsin-Madison in 1975 and from Harvard Law School (with honors) in 1978. He became a trial lawyer because discovering the truth about things has always been very important to him. He litigated employment discrimination cases all over the country for about twenty years before retiring to West Virginia where his wife and he homeschooled their twin daughters through their completion of high school.

His passion for the topic of evolution comes from personal experience. Bob was trained to believe that Darwin's theory was as well-established as the theory of gravity. In all his years of schooling, he never heard a single word of criticism or doubt about Darwin's big idea. But when he became a father of precious twin daughters in 1992 at almost forty years of age, he decided to study the topic for himself so he could teach his daughters the truth. When he began to discover the truth about evolution, it changed his life completely because his materialistic worldview about the origin and nature of humans no longer had a credible basis.

His study of the topic of evolution continued for more than twenty-five years. Over the years, he digested a wide variety of books on the subject that often required a great deal of study. In 2020, he was moved to write his own book on the topic because he realized that what people think about where humans came from has a huge impact on how people think about life, yet almost no one has the time to study the evidence the way he had done. So he distilled decades of study and explained in simple terms why Darwin's big idea cannot be true. His book enables the reader to understand in a matter of hours what has taken him years of study to comprehend.

Bob's training and experience as an attorney in proving and disproving facts informs the first chapter of the book which, among other things, helps readers understand how to distinguish evidence from imagination and opinion, and how to assess the credibility of evidence. Subsequent chapters present the case against Darwinism on several dimensions. In addition to an overview of the fossil evidence, these include (a) the logical impossibility of highly complex biological systems being created by mutation and natural selection, (b) the fact that not even an ancient universe would have allowed enough time for evolution to have occurred, and (c) the shocking reality that each of the "icons of evolution" presented in high school and college biology textbooks has actually been debunked or, in some cases, proven to have been a fraud in the first place.

www.ingramcontent.com/pod-product-compliance
Lightning Source LLC
Chambersburg PA
CBHW040108180526
45172CB00009B/1266